T0314536

Night Vision

Drawing on exciting discoveries of the last 40 years, *Night Vision*
explores how infrared astronomy, an essential tool for modern
astrophysics and cosmology, helps astronomers reveal our universe's
most fascinating phenomena – from the birth of stars in dense
clouds of gas to black holes and distant colliding galaxies, and the
cycle of interstellar dust from its origin in outflows from dying
stars to the formation of our solar system. While surveying the
progress in infrared observation and theory, astronomer Michael
Rowan-Robinson introduces readers to the pioneering scientists and
engineers who painstakingly developed infrared astronomy over the
past 200 years. Accessible and well illustrated, this comprehensive
volume is written for the interested science reader, amateur
astronomer, or university student, while researchers in astronomy
and the history of science will find Rowan-Robinson's detailed notes
and references a valuable resource.

MICHAEL ROWAN-ROBINSON served as Head of Astrophysics at Imperial
College London from 1993 to 2007 and as president of the Royal
Astronomical Society from 2006 to 2008. He has received several
awards for his work in infrared and submillimetre astronomy,
including the first Institute of Physics Hoyle Medal in 2008. He
worked on the *IRAS*, *ISO*, *Spitzer*, and *Herschel* missions, and was
involved in the pioneering submillimetre observations of the 1970s.

Night Vision

Exploring the Infrared Universe

MICHAEL ROWAN-ROBINSON

Imperial College London

CAMBRIDGE UNIVERSITY PRESS
Cambridge, New York, Melbourne, Madrid, Cape Town,
Singapore, São Paulo, Delhi, Mexico City

Cambridge University Press
32 Avenue of the Americas, New York, NY 10013-2473, USA

www.cambridge.org
Information on this title: www.cambridge.org/9781107024762

First published 2013

Printed in the United States of America

A catalog record for this publication is available from the British Library.

Library of Congress Cataloging in Publication Data
Rowan-Robinson, Michael.
  Night vision : exploring the infrared universe / Michael Rowan-Robinson.
    pages   cm
  Includes bibliographical references and indexes.
  ISBN 978-1-107-02476-2
  1. Infrared astronomy. I. Title.
  QB470.R69   2012
  522'.683–dc23   2012005511

ISBN 978-1-107-02476-2 Hardback

For Artie, Amelie, Alice and Lizzie

Contents

Color plates follow page 118

Preface

It has been a huge challenge to try to write the whole story of infrared astronomy, from the discovery of infrared radiation from the Sun by William Herschel (1738–1822) in 1800 through to the present day, to the discoveries being made by the space mission named after Herschel. I wanted to make this story accessible to the general reader with some interest in science but no scientific background. At the same time I wanted to make it a full and accurate account. Having lived and worked through the great period of infrared astronomy, from the 1960s to the present, I know many of the major figures whose work is described here, and I wanted to do them justice.

To reach the general reader, I have had to continually simplify the text, moving more complex and detailed material to the notes. Astronomy is a branch of physics, and physics is not an easy subject for someone who has perhaps not even studied it at school. I've provided a glossary of technical terms and tried to keep them to a minimum. The notes and very full bibliography allow the interested reader to explore the full details of a major area of science.

I have written about what I know, the science of infrared astronomy, and haven't attempted to give the full story of the technological developments required to make this science possible. To give some idea of the huge army of people who work to provide the tools for science, an infrared space mission generally involves more than one thousand people in its design and construction. I was drawn into infrared astronomy in the early 1970s by my friend, and colleague at Queen Mary College London, Peter Clegg. It was through him that I became involved in the first infrared space mission, the *Infrared Astronomical Satellite* (*IRAS*), which so transformed infrared astronomy. Consequently, my career in infrared astronomy, and this book, are very much thanks to Peter.

I cannot begin this account of infrared astronomy without mentioning the book by my late friend David Allen, *Infrared, the New Astronomy*, published in 1975.[1] David was a student contemporary of mine, and I met him several times in those days. He and another student friend, Michael Penston, also sadly deceased, ran the student astronomical society. Both became distinguished astronomers; David ended up as a senior figure at the Anglo-Australian Observatory, based in Sydney. He specialized in infrared astronomy and built infrared instruments for the Anglo-Australian Telescope, a 4.2-metre-diameter telescope built jointly by Australia and Britain. Later I found myself chair of the Anglo-Australian Telescope Board, which managed the telescope and observatory, at a time when David was terminally ill with cancer. Shortly before his death, I went with the members of the Board to visit him at his bedside and consult him about the future of the observatory.

Infrared, the New Astronomy is a delightful and idiosyncratic account of the emergence and early days of infrared astronomy. There are dark hints of rivalries and disputes between the early pioneers. This was before the launch of *IRAS*, and there are only the first glimpses of the new astronomy that was emerging. But it is fascinating to read this book to get a very different perspective from the one we have today.

More authoritative reviews of the history of infrared astronomy, and of the early days of modern infrared astronomy, have been given by Frank Low, George Rieke, and Robert Gehrz, and I have used these very freely.[2,3] I also found Malcolm Longair's magnificent survey of the history of modern astronomy, *The Cosmic Century*, essential reading.[4] There have been more than 30 reviews of different aspects of infrared astronomy in the excellent series *Annual Review of Astronomy and Astrophysics*, which I found invaluable.

I'd like to thank several colleagues for supplying material, reading the manuscript, correcting my errors, and making very helpful suggestions: Mike Hauser, Mike Werner, Tom Phillips, George Rieke, Ian Robson, George Helou, Ian McLean, Steve Price, Peter Ade and Matt Griffin. I thank John Herschel-Shorland for showing me his archive of William Herschel material and for letting me use his portrait of William Herschel. The late Peter Hingley was, as always, helpful in locating historical material and images. Several people very kindly played the role of the general reader and helped me towards a more comprehensible book: Stephen Curtis, the late David Marcus, and above all, my wife, Mary.

1

Introduction

Night vision – the ability of infrared light to penetrate dust and to light up the dusty universe – and also the new vision of the night sky, the universe that infrared astronomy has unveiled.

All astronomy starts from the night sky, the picture of the universe we see with our own eyes: the stars and constellations, the Milky Way. We have become used to the idea that there is a universe beyond the constellations, revealed by the giant telescopes of the astronomers. We can still imagine the *Hubble Space Telescope* to be a giant extension of our own eyes, and its images are still images made in visible light. Much harder to grasp is the universe that is revealed in invisible light, such as the infrared radiation that is the subject of this book. I've called the book *Night Vision* for two reasons. Firstly, infrared sensors and binoculars are already widely used to aid seeing in the dark, or night vision. And secondly, my goal is to try to make the infrared universe as familiar to you as the night sky, so that when you look out at the night sky you can also imagine what it would look like with infrared eyes, and therefore see the new vision of the night sky that the infrared gives us.

Above all, infrared astronomy is about the cool, dusty universe. Spread between the stars are tiny grains of dust, similar to sand and soot, and these absorb the light from stars and reradiate the energy as infrared radiation. There are dense clouds of gas and dust within which new stars are forming, and only with infrared light can we peer into them. Dying stars and massive black holes are often shrouded in dust, which shines in the infrared. And the cool bodies of the Solar System, planets, comets and asteroids are mainly radiating infrared light. In the infrared and submillimetre parts of the spectrum, we see galaxies in formation, undergoing violent bouts of star formation often caused by collisions between smaller fragments. And we see the cool glow left over by the hot fireball phase of the Big Bang itself.

NIGHT VISION

There are two subtle things about human night vision. Firstly, it takes about half an hour to get fully working, and we need to be away from any direct light. Initially, if we go outside at night from a lighted room, we blink and see very little of the stars. After a while, the eye adjusts and we start to see the fainter stars. And secondly, our night vision works better at the edge of our visual field than in the centre. To see the faintest stars and nebulae, we need to look 'out of the corner of our eye'.

Now these are consequences of the way the human eye works. There are two types of receptors in the eye, *rods* and *cones*. The cones are packed into the centre of the eye's detector, the retina, and are responsible for our main vision. They are sensitive to colour, and there are three distinct types of cone receptors, with sensitivity to red, green or blue light. By combining the information from these three types of cones, the brain is able to give us the full experience of colour, and we are able to discriminate more than one hundred different shades of colour. The far more numerous rods, by contrast, are dominant at the edge of the retina and are more sensitive to low levels of illumination, hence their importance for night vision. They do not have any colour discrimination, but they are very sensitive to movement, and so we catch sight of something moving behind us at the edge of our vision. It's as if the eye is really two visual systems in one, one for normal daytime vision and the other for night vision and for detecting movement.

If we go out into a dark night, then at first our vision is very insensitive. Within about five minutes we notice a dramatic improvement in sensitivity – by a factor of 50. This results from the cones at the centre of our vision becoming adapted to the dark. After about half an hour there is an even bigger improvement, by a further factor of 200, as the rods become adapted. In other words, from first entering the dark to full dark adaptation half an hour later, the sensitivity of our vision has improved by a factor of 10,000.

When our eyes have become 'dark adapted', on a clear night, away from the polluting light of towns and cities, we then have the wonderful display of the night sky: the sparkling stars with their different colours, red for Betelgeuse, the Armpit of the Giant, Orion; blue for Vega, the brightest star in Lyra; and white for Sirius, the Dog Star. We see star clusters such as the Pleiades, and the familiar constellations of Ursa Major, Cassiopeia and Orion that we can all recognize, which are delineated by bright stars that happen to lie together on the sky but can be at a wide range of distances. And we can pick out the wandering planets and the Moon, and the fuzzy stream of light across the sky, the Milky Way, which is the flattened distribution of hundreds of billions of stars of our own Galaxy, seen from the Solar System's location near the edge. All this looks even better if you can get to a mountaintop, as astronomers like to do. And then with our wonderful modern telescopes we open up a more distant landscape of billions of galaxies and reveal shoals of faint stars and hot gas clouds in our Milky Way Galaxy, often streaked with dark patches.

I want to try to convince you that this familiar night sky and the universe of visible galaxies beyond are only a small fraction of the cosmic world. About half of the light that reaches the Earth from the universe is invisible to the human eye and is infrared light. And the infrared cosmic landscape is almost unimaginably different from the night sky we see with the human eye or with optical telescopes. It has only been during the past 50 years that the veil on this cool, dark landscape has been drawn back. I've been lucky enough to know most of the pioneers of modern infrared astronomy and to participate in some of these great discoveries.

Across the various species, eyes appear to have evolved independently several times, and Simon Conway Morris, in his book *Life's Solution*, emphasizes how some of these different eyes, for example those of the octopus and humans, have converged to a rather similar design.[1] Especially interesting is the case of the rattlesnake, which has evolved a primitive infrared eye. This allows it to see in the dark, where its prey appears as a warm, glowing, bright infrared object (see Figure 1.1). So the rattlesnake really does have night vision. If humans had evolved this ability, I would not need to be writing this book. I shall try to bring to life, for infrared-blind humans, the invisible universe.

Because we do not have infrared vision, we have had to develop ways of detecting infrared radiation, and until recently this was a slow process. Today soldiers routinely have infrared binoculars to scan the night-time battlefield. Firefighters use infrared goggles to detect people in smoke-filled buildings. And wildlife programmes on television use infrared cameras to watch nocturnal animals such as foxes and badgers. Infrared binoculars were first developed for military use during the Korean War, but much of the routine infrared capability we now use, including our television remote control, has developed only in the past 20 years.

Infrared radiation was first discovered by William Herschel in 1800. Isaac Newton (1642–1727) had shown that if the light from the Sun is passed through a prism, it becomes spread out into the colours of the rainbow, the *spectrum*. For mystical reasons, he chose to characterize this as seven colours: red, orange, yellow, green, blue, indigo and violet. In reality, the colours merge imperceptibly from one to another and there are an infinite number of possible colours. Herschel asked himself the question: what happens beyond the red and beyond the violet? By placing a mercury bulb thermometer beyond the red end of the spectrum and watching the mercury rise up the tube of the thermometer, he showed that the Sun is emitting radiation beyond the red end of the spectrum, infrared radiation. And he realized that this radiation, which he called 'invisible light', is the same thing as radiant heat. Now we do not 'see' radiant heat but we can feel it through our skin, so our skin is a detector of infrared radiation. You will certainly have noticed that when many people gather in a cold room or railway carriage, it soon warms up from the heat of the bodies present. The human body radiates about a hundred watts of infrared power, so our body's ability to sense infrared radiation is our third, crude eye.

Newton thought that light consisted of 'corpuscles', or discrete particles. Today we know this is partly true, although Newton had the completely wrong idea that these particles of light were emitted by the eye. Through the eighteenth century it began to be realized, by Robert Hooke (1635–1703), Christiaan Huygens (1629–1695), Leonhard Euler (1707–1783) and especially Thomas Young (1773–1829), that the best way to understand most of the known phenomena of light was to think of light as waves. The different colours then corresponded to different wavelengths of light. These waves are very short in length and have to be measured in terms of microns, a micron being a millionth of a metre, or a thousandth of a millimetre. To set the scale, a human hair is about 100 microns in diameter. Red light corresponds to light of wavelength 0.7 microns and violet light has wavelength 0.4 microns, so the visible spectrum of light spans the narrow range of wavelengths from 0.4 to 0.7 microns. What Herschel had shown was that there is radiation with a wavelength longer than 0.7 microns. Only a few years later, Johann Ritter (1776–1810) showed the existence of ultraviolet

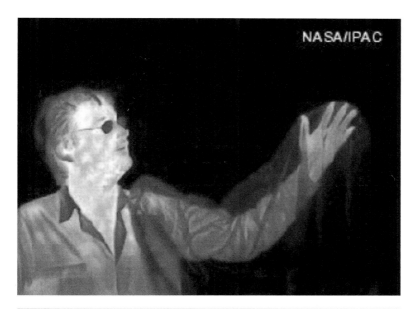

Figure 1.1. The same scene viewed in visible and infrared light. In visible light the arm is concealed by the black polythene sack, but it can be seen clearly in infrared light.

radiation, light with a wavelength less than 0.4 microns, by showing that silver chloride was blackened by light from the Sun beyond the violet end of the spectrum.

The breakthrough in understanding came when the Scottish physicist James Clerk Maxwell (1831–1879) realized in 1864 that light consists of vibrations in a combined electric and magnetic field.[2] By uniting and extending the laws of electricity and magnetism that had been developed by Carl Friedrich Gauss (1777–1855), Michael Faraday (1791–1867) and André-Marie Ampere (1775–1836), Maxwell was able to show that there could be electromagnetic waves and that their speed would be exactly the same as the known speed of light. These waves could be characterized by their wavelengths, and the difference between the then known types of light, namely visible, infrared and ultraviolet, was simply one of wavelength. The wavelength is simply the distance between successive crests or successive troughs of the wave. The type of wave can also be characterized by the number of waves passing per second, or frequency. The product of the wavelength and frequency is then just the speed of light, which is the same for all types of light. At any point in space, an electric charge will be pushed in the direction of the local electric field, if there is one. A magnet will line itself up in the direction of the prevailing magnetic field. When a light wave passes, what really happens is a trembling of the electric and magnetic fields and then stillness again. That trembling is the light wave, so we now talk about the *electromagnetic spectrum* of radiation, meaning all the possible kinds of light or radiation. Starting at the shortest wavelengths, we have gamma rays, x-rays, ultraviolet, the visible range from violet to red, then infrared, submillimetre, microwave and radio. Maxwell's prediction of electromagnetic radiation of different wavelengths was confirmed by the discovery of radio waves by Heinrich Hertz (1857–1894) in 1887 and of x-rays by Wilhelm Roentgen (1845–1923) in 1895.

While Maxwell's theory brilliantly explained the known properties of light, the advent of quantum theory in the early twentieth century introduced another aspect of light. In 1905, Albert Einstein (1879–1955) showed that we have to think of light both as a wave and as a particle. The particle of light became known as the *photon*, and a photon carries a precise amount of energy, which is inversely proportional to the wavelength of the light. So x-ray photons have very high energy, whereas radio photons have low energy. From gamma rays to the longest-wavelength radio waves there is a factor of one thousand million million (10^{15}) increase in wavelength (Figure 1.2). The visible

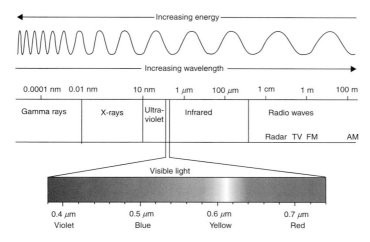

Figure 1.2. The electromagnetic spectrum, showing the gamma ray, x-ray, ultraviolet, visible, infrared and radio bands. The visible band stretches from wavelengths 0.4 to 0.7 microns, the infrared and submillimetre band from 0.7 to 1000 microns (1 millimetre).

band is just a factor two spread in wavelength, only one-fiftieth of the total. It's amazing what a world of colour we experience from this tiny range of wavelengths. But equally the universe of invisible wavelengths is unimaginably different from the familiar universe of the visible band.

The twentieth century saw the opening up of all these different wavebands for astronomy and the birth of the astronomy of the invisible wavelengths. Radio astronomy began quietly with an investigation of radio "static" interference at a wavelength of 15 metres by Karl Jansky (1905–1950) at Bell Telephone Laboratories in Holmdel, New Jersey, in 1930–3. He built a rotatable aerial array 30 metres long and 4 metres high, which was nicknamed the 'merry-go-round', mounted on four wheels taken from a Model T Ford (Figure 1.3). He found that part of the static was caused by thunderstorms, but there remained a steady hiss from a direction that moved around the sky a little each day. This direction turned out to be the constellation Sagittarius. Jansky had discovered radio emission from the Milky Way. This discovery had little media impact apart from a sardonic comment by *The New Yorker*: 'This is the longest distance anyone ever went looking for trouble'. Few professional astronomers took any notice of Jansky's results either, and the next step was taken by an American amateur astronomer, Grote Reber (1911–2002), who made detailed radio maps of the Milky Way between

1938 and 1944. Meanwhile, in wartime Britain, John Hey (1909–2000), investigating apparent enemy jamming of radar installations, discovered radio emission from the Sun. After the war, he went on to detect several discrete astronomical sources of radio waves and inspired the formation of radio astronomy groups at Cambridge and Jodrell Bank in Britain, and in Australia. The big breakthrough came with radio surveys of the sky made in the 1950s and 1960s, especially at Cambridge and in Australia. These led to the discovery of remarkable new types of objects, galaxies that were very powerful emitters of radio waves, called *radio galaxies*, and distant radio sources which looked like stars at optical wavelengths, which became known as quasi-stellar radio sources, or *quasars*. The latter were hundreds of times more luminous than the light from our whole Galaxy, yet their rapid variability, on timescales of years or months, meant that this huge output was coming from a region not much larger than the Solar System. Ultimately these regions were understood as massive black holes at the centres of galaxies. A *black hole* is formed when a massive body collapses under the influence of gravity into such a compressed state that light can no longer escape from it. The black holes responsible for quasars are typically a hundred million times more massive than the Sun. The 1960s saw the discovery of another exotic type of object, *pulsars*, radio sources that pulsate rapidly and are associated with rapidly spinning compact neutron stars. A *neutron star* is a dead star that has exhausted its nuclear fuel and become so compressed that its radius is only ten kilometres, smaller than Los Angeles or London.

While radio waves from the universe reach the ground, x-ray astronomy is only possible from space, because the Earth's atmosphere strongly absorbs x-rays. X-ray astronomy began in 1948 when American Thomas Burnight detected x-rays from the Sun by using a captured German V2 rocket. In 1962 Riccardo Giaconni discovered a compact x-ray source in the constellation of Scorpius while trying to observe the Moon with a rocket-borne x-ray detector. This was the first of many x-ray sources found that are associated with very compact dead stars. When a star reaches the end of its life, its core collapses to form a very small remnant: a star like the Sun ends up as a *white dwarf* star, a very hot object about the size of the Moon. More massive stars end up either as neutron stars or the ultimate compact object, a black hole. The x-ray emission comes from gas that falls onto the compact object or forms a disk of very hot gas orbiting it. The first x-ray survey of the sky was undertaken by NASA's *Uhuru* satellite in 1970, and there have been many subsequent x-ray missions of

Figure 1.3. The 'merry-go-round' antenna with which Karl Jansky, pictured here, discovered radio emission from the Milky Way in 1930–3.

ever-increasing sophistication. *Uhuru* found dozens of compact x-ray sources, the first x-ray galaxies and quasars, and x-ray emission from very hot (ten million degrees Kelvin) gas in clusters of galaxies, and detected x-ray background radiation from sources spread through the whole universe.

In this book we are going to focus on infrared and submilli-metre radiation, light with wavelengths from 0.7 microns to 1 mil-limetre (1000 microns). Because this is a rather broad swathe of wavelengths, we often subdivide it into near infrared (0.7–3 microns), mid-infrared (3–30 microns), far infrared (30–200 microns) and sub-millimetre (200 microns to 1 millimetre). For reasons having to do with detector technology, we often think of the submillimetre band as extending to 3 millimetres, although 1–3 millimetres cannot really be "sub"millimetre. For astronomy what unites the whole infrared and submillimetre band is the phenomenon of interstellar dust, which absorbs visible and ultraviolet light from stars, galaxies and hot disks around black holes and then reradiates it at infrared and submillime-tre wavelengths. When we look up at the Milky Way at night, we see

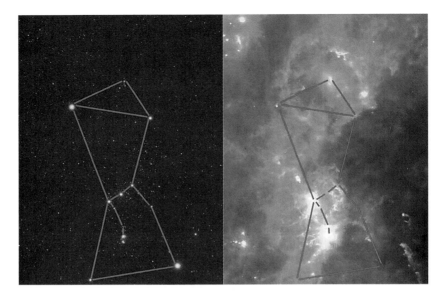

Figure 1.4. On the right is an infrared image of the constellation Orion, from the *Infrared Astronomical Satellite* mission (see Chapter 7), compared with an optical image on the left. The lower of the two bright patches in the lower centre of the infrared image is the Orion Nebula (Messier 42) in the Sword of Orion, while the bright nebulosity just above it and to the left surrounds the Belt star Zeta Orionis. The bright spot surrounded by a large ring is Lambda Orionis, and the spot just outside this ring on the left is Betelgeuse (see also Plate I).

that parts of it are obscured by dark patches such as the Coal Sack, which are in fact clouds of gas and dust. In infrared light these same dark clouds appear as bright patches of emission. In fact, about half the energy in starlight ever emitted in the universe has been absorbed by dust and reradiated in the infrared. For newly formed stars that have not yet emerged from the cocoon of gas and dust in which they formed, the fraction of their visible and ultraviolet light absorbed by dust can exceed 99%. At infrared wavelengths, we can peer into these obscured regions, so there are aspects of our world that simply cannot be appreciated without looking in infrared light. This is wonderfully illustrated by the contrasting infrared and optical views of the constellation Orion (Figure 1.4, see also Plate I). The bright stars of the visible constellation are barely discernible in the infrared. Instead we see the clouds of dust and gas from which new stars are forming.

THE COOL UNIVERSE

Above all, at infrared wavelengths we are studying the cool universe, compared with the hot surfaces of stars that we see in visible light and the very hot gas that we see at x-ray wavelengths. Physicists and astronomers like to use a scale of temperature invented by Lord Kelvin in 1848 which is similar to the Celsius (C) scale but starts at the absolute zero of temperature, −273 °C. So ice melts at 273 K (degrees Kelvin), and water boils at 373 K. Nitrogen liquefies at 77 K (−196 °C) and helium at 4.2 K. At 0 K (−273 °C) the random motions of all atoms and molecules cease, and it is impossible to make a body colder than this.

In 1900, Max Planck (1858–1947), the founder of quantum theory, discovered that for a perfectly efficient absorber or emitter of radiation, which is known as a *blackbody*, the distribution of brightness with wavelength, or spectrum, has a very characteristic shape, which depends only on the temperature (Figure 1.5). We see a Planck blackbody spectrum when we peep into an oven or furnace, because in such a cavity the walls of the cavity and the radiation reach a state of thermal equilibrium. The temperature of the radiation becomes equal to the temperature of the matter it is in contact with. For a blackbody there is a direct proportionality between the temperature and the frequency of peak intensity (so there is an inverse relationship between temperature and wavelength). The surface of the Sun looks yellow, corresponding approximately to a blackbody of 5800 K. The red star Betelgeuse in Orion has a surface temperature of about 3000 K. The blue star Spica is at 25,000 K. For x-ray radiation, we are talking about temperatures of 1 million to 100 million degrees Kelvin. With infrared and submillimetre radiation, we are studying much cooler temperatures, in the range 3 K (at a wavelength of 1 millimetre) to 1000 K (at a wavelength of 3 microns). This includes planets, asteroids, comets, brown dwarfs, newly forming stars and galaxies, dust shells around dying stars, dusty disks around stars where planets are being formed, and the dense clouds of gas and dust where new stars are being born. The average temperature of the Earth is 288 K (15 °C), so the main radiation from the Earth is at infrared wavelengths. Jupiter is at 150 K. The dust around dying stars can be at 100–1000 K and dense star-forming clouds are at about 50 K, while the dust clouds in the Milky Way are at 10–30 K. The cosmic background radiation left over from the Big Bang has a temperature of about 3 K.

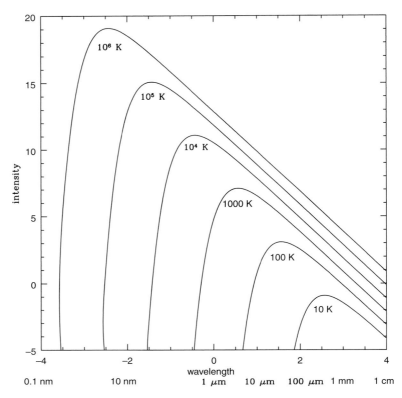

Figure 1.5. Planck blackbody curves, intensity versus wavelength, for temperatures 10 to 1,000,000 K, illustrating how the peak wavelength depends on temperature. Cold dust, at 10 K, peaks at submillimetre wavelengths. Hot dust, at 1000 K, peaks at near-infrared wavelengths. A hot star, at 10,000 K, peaks at ultraviolet wavelengths. And very hot (1 million degrees Kelvin) gas peaks at x-ray wavelengths.

THE SLOW DEVELOPMENT OF INFRARED ASTRONOMY

Why did it take so long for infrared and submillimetre astronomy to develop? William Herschel discovered infrared radiation in 1800, but progress was extremely slow for the next 150 years, and the real explosion in infrared astronomy did not happen until the 1980s, when the *Infrared Astronomical Satellite* (IRAS) surveyed the sky. There are three main reasons for this. Firstly, the technology was slow to develop and it took almost a century for detectors capable of recording the faint radiation from stars to be invented. Secondly, the sky is extremely bright in the infrared because of the emission from the Earth's

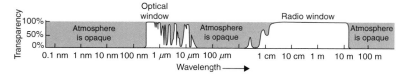

Figure 1.6. Atmospheric transmission at sea level across the
electromagnetic spectrum. Apart from windows in the optical,
near-infrared, millimetre and radio bands, the Earth's atmosphere
is opaque. Some additional windows open up at submillimetre
wavelengths from high mountaintops, but otherwise we have to go to
space to see the full range of wavelengths.

atmosphere. It's a bit like trying to observe the stars in visible light in
the daytime: the stars are certainly there, but their light is drowned
out by scattered sunlight. In the infrared, we have to use differencing
techniques, where we look for the difference in the brightnesses of
two nearby regions of the sky. Hopefully the emission from the atmosphere will be the same and the difference will be the light from the
star. Finally, there is the very strong absorbing effect of the Earth's
atmosphere (Figure 1.6). Very little of the infrared and submillimetre
radiation from stars, galaxies or the universe reaches the ground. The
reason for this is that molecules such as water and carbon dioxide are
strong absorbers of infrared radiation. These molecules radiate and
absorb at a multitude of infrared wavelengths, which overlap strongly
to create a dense infrared fog. There are just a few narrow wavelength
windows, at 1.25, 1.65, 2.2, 3.5, 5, 10 and 20 microns in the near and
mid-infrared and at 350, 450 and 850 microns in the submillimetre,
through which astronomers can work from mountaintop sites. To see
the whole infrared waveband, we have to go to space, and that was not
possible until the 1980s, though heroic efforts to get above the atmosphere were made from aircraft, rockets and balloons in the 1970s.

The absorbing effects of the Earth's atmosphere are important
for human survival because they are responsible for the 'greenhouse'
effect. Radiation from the Sun at visible wavelengths is absorbed by
the Earth and reradiated at infrared wavelengths. But this infrared
radiation cannot properly escape through the molecular blanket of
the Earth's atmosphere so the Earth heats up. If there were no atmosphere, the mean temperature of the Earth would be −18 degrees
Celsius. The greenhouse effect warms the Earth to a mean temperature of about 15 degrees Celsius. The main absorbing gases are carbon
dioxide, which dominates the absorption from 1 to 5 microns, and

water vapour, which dominates from 5 microns to 3 millimetres. The peak wavelength of the Earth's emission is at about 10 microns, so water vapour is the dominant 'greenhouse' gas, with carbon dioxide contributing about 20% of the greenhouse effect.

In this book I will first describe how William Herschel discovered infrared radiation and launched the astronomy of the invisible wavelengths. I will outline the slow progress that was made over the next 150 years and then how systematic studies of dying stars and stars being born developed in the 1950s and 1960s. Then I tell the story of the birth of far-infrared and submillimetre astronomy in pioneering experiments from aircraft, balloons and rockets, and the discovery of the cosmic microwave background radiation. The launch of the *Infrared Astronomical Satellite* (*IRAS*) in 1983 was a turning point for infrared astronomy and was followed by space missions dedicated to studying the cosmic microwave background, *COBE* and *WMAP*. Throughout I will emphasize the importance of theoretical work in trying to understand the infrared universe that observations were revealing. I will discuss the impact of the giant ground-based infrared and submillimetre telescopes, and the dedicated infrared space observatories such as *ISO* and *Spitzer*. I then turn to our dusty solar system and talk about the zodiacal dust, comets and asteroids, and planets around other stars (known as exoplanets). Finally, I talk about the new space missions launched in 2009, *Herschel* and *Planck*, and the other exciting new infrared and submillimetre facilities planned for the next decade.

2

William Herschel Opens Up the Invisible Universe

William Herschel (Figure 2.1, Plate II) was one of the greatest astronomers of all time, certainly the greatest since Galileo. His achievements dominated his age in a way that did not happen in the twentieth century, even though there have been some great astronomers in our time. Herschel discovered the first new planet in recorded human history, Uranus; constructed telescopes of a size and quality which were unrivalled in his day; demonstrated the disk-shaped structure of our Milky Way Galaxy; and surveyed the whole northern sky for nebulae and demonstrated that many were star clusters. He showed that the universe was enormously larger than had previously been imagined and introduced the idea that stars are born and die. He thought that some nebulae might be distant star systems similar to the Milky Way, but later doubted this. But above all Herschel discovered that the Sun emits radiation beyond the red end of the spectrum – infrared radiation – and thereby opened up the astronomy of the invisible wavelengths that has been a dominant feature of astronomy since 1960.

Remarkably Herschel was an amateur, self-taught astronomer, and might have remained an obscure amateur but for his discovery of Uranus. His father was a bandsman in the Hanoverian army who also loved astronomy and who imparted his love of music and astronomy to his children.[1] In due course, at the age of fourteen, William also became a bandsman, and father and son later campaigned with the Hanoverian Foot-Guards in the War of the Austrian Succession. Following the French occupation of Hanover, Herschel's parents were concerned about William's safety, so in 1757 they got him out of the army and sent him to England, at age eighteen, to seek his fortune as a musician. He first worked copying music, then as bandmaster of a regimental militia in Yorkshire, and then for many years as a freelance

Figure 2.1. Sir William Herschel.

performer and composer in the north of England. After ten years of this unsettled existence, he became organist of the Octagon Chapel in Bath in 1767, earning a comfortable income. He brought his sister Caroline from Hanover to keep house and launch her on a musical career as a singer. Meanwhile William had become even more interested in astronomy and had begun to construct his own telescopes. He started a series of ever deeper, systematic surveys, searching for double star systems. It was during his second survey, with a 7-foot-long reflecting telescope, that on 13 March 1781 he made a discovery that was unprecedented in human history, a new major planet of the Solar System. Since antiquity the known planets had been Mercury, Venus, Mars, Jupiter and Saturn, all visible to the naked eye. But Herschel's discovery was no lucky accident. He was engaged in a survey of the whole northern sky and was bound to find it if it was there. For this discovery Herschel was awarded the Copley Medal of the Royal Society and elected a fellow of the Society.

The discovery of Uranus also brought William Herschel to the attention of King George III, not least because Herschel offered to name the planet after him. The king provided Herschel with a salary so that he could settle in the neighbourhood of Windsor and devote

Figure 2.2. Herschel's 40-foot telescope, with which he demonstrated the vast depths of the universe and the birth and death of stars.

himself to astronomy, with the requirement that he should be available to show the Royal Family celestial objects of interest through the telescope. He was able to devote himself to his surveys of the sky for nebulae and star clusters, assisted by his sister Caroline. In 1788 he married, and in 1792 his son John (1792–1871), who became a distinguished astronomer and mathematician as well as a pioneer of photography, was born.

William Herschel was the first modern astronomer. His 40-foot telescope (Figure 2.2), with a mirror diameter of 4 feet (1.2 metres), was a forerunner of giant modern reflecting telescopes. He used it for a series of systematic studies of the main types of unusual celestial objects known in his day: variable stars, double stars and nebulae. The 40-foot telescope was extremely difficult to use, however, and Herschel used his other smaller telescopes for much of his work. He initiated survey astronomy, which has been at the heart of astronomical research ever since. His catalogue of nebulae, extended into the Southern Hemisphere by his son John, became the basis of the *New General Catalogue of Nebulae and Star Clusters*, which is still in use today.

William Herschel's studies of nebulae were continued by William Parsons (1800–1867), Lord Rosse, who constructed an even larger telescope, the 6-foot diameter 'Leviathan' in 1848, and used it to show that some of the nebulae showed spiral structure. Herschel's surveys opened up the concept of the depths of the universe, the vast scale of the Galaxy,[2] and the huge distances of other galaxies. He also introduced the idea that stars were forming out of clouds of gas, that the whole stellar universe is evolving.[3,4]

It was a remarkable set of achievements for someone whose only education had been in the Garrison School in Hanover. In 1820, a group of astronomers, including Herschel's distinguished son John, wanted to start an astronomical society in London. The usual practise for scientific societies was to find an interested nobleman to be their president, and they approached the Duke of Somerset. However, Joseph Banks, president of the Royal Society, did not want this new astronomical society to be formed, fearing this would undermine the Royal Society. He persuaded the Duke of Somerset to withdraw his name. The astronomers then invited a far more distinguished figure to become their president, William Herschel. By this time Herschel was not in good health, and he never attended a meeting of the Society. But the Royal Astronomical Society, as the Society became, is immensely proud that William Herschel was its first president. The first honorary member of the Society was Caroline Herschel, who had contributed so much to her brother's scientific programme and made many discoveries in her own right.

It was while studying sunspots and solar granularity in the 1790s that William Herschel became interested in the possibility that light of different colours might have different heating powers. To observe the Sun, he had to protect his eyes with coloured glass shades, and he noticed that the sensation of heat seemed to vary according to the colour of the glass:

> In a variety of experiments I have occasionally made, relating to
> the method of viewing the Sun, with large telescopes, to the best
> advantage, I used various combinations of differently-coloured glasses.
> What appeared remarkable was, that when I used some of them, I felt
> a sensation of heat, though I had but little light; while others gave
> me much light, with scarcely any sensation of heat. Now as in these
> different combinations the Sun's image was also differently coloured,
> it occurred to me, that the prismatic rays might have the power of
> heating bodies very unequally distributed among them; and, as I judged
> it right in this respect to entertain a doubt, it appeared equally proper
> to admit the same with regard to light.[5]

To investigate this, in 1800 Herschel set up an experiment where the light from the Sun was passed through a narrow slit and then a prism to make a spectrum of the colours from red to violet. He placed a thermometer in the red part of the spectrum and recorded the rise in temperature compared with another thermometer close to the first but in the shade. The difference in readings on the thermometers was assumed to be an indication of the heating effect of the Sun's red light. Herschel then repeated this for green and violet light. The average heating effects were found to be proportional to 55, 24 and 16, respectively, for red, green and violet light. Herschel next compared the illuminating powers of various coloured lights by examining through a microscope objects illuminated by different colours. He found that for the human eye, 'the maximum of illumination lies in the brightest yellow, or palest green'. Herschel suspected that the heating effect of sunlight did not stop at the red end of the visible spectrum:

> I must now remark, that my foregoing experiments ascertain beyond a doubt, that radiant heat, as well as light, whether they be the same or different agents, is not only refrangible, but is subject to the laws of the dispersion arising from its different refrangibility.... May this not lead us to surmise, that radiant heat consists of particles of light of a certain range of momenta, and which range may extend a little farther, on each side of refrangibility, than that of light? We have shewn, that in a gradual exposure of the thermometer to the rays of the prismatic spectrum, beginning from the violet, we come to the maximum of the light, long before that of heat, which lies at the other extreme....
>
> [T]he full red falls still short of the maximum heat; which perhaps lies a little beyond visible radiation. In this case, radiant heat will at least partly, if not chiefly, consist, if I may be permitted the expression, of invisible light; that is to say, of rays coming from the Sun, that have such a momentum as to be unfit for vision.

A month later he described an experiment to test this. He now had three thermometers, one of which was placed in line with the solar spectrum but at different distances beyond the red end of the visible spectrum. The two others were placed in line with the first but off to the side, out of the sunlight. The exposed thermometer showed a higher temperature than the two others, demonstrating that there were heating rays beyond the red end of the spectrum (Figure 2.3, Plate III). The maximum heating effect occurred about half an inch beyond the red end of the spectrum, with a heating effect still being detected an inch and a half beyond the red end, compared with the eight-inch length of his visible, violet to red, spectrum.[6] This suggests his measurements extended to a wavelength of about 0.8–0.9 microns. Herschel's experiments showed that radiant heat, like light, is subject to refraction because the prism dispersed it. In two further substantial papers, he describes 219 experiments in which he goes on to show that the heating rays, whether solar or terrestrial in origin, obeyed the same laws of reflection and refraction as visible light.[7] Herschel's initial view had been that light and radiant heat were essentially identical, but he wavered in this correct view following some of his later experiments on the absorption of heat and light by transparent media.[8] It was not until Maxwell's 1864 demonstration that light consisted of electromagnetic radiation that the common nature of visible and infrared radiation was conclusively demonstrated.

Herschel had also tried his experiment with thermometers in the region beyond the violet end of the spectrum, but found no effect. In 1803, the German physiologist Johann Wilhelm Ritter, possibly the model for Mary Shelley's Dr. Frankenstein,[9] demonstrated the existence of invisible rays beyond the violet end of the spectrum, ultraviolet light, through the blackening effect on silver chloride, which is the basis of the photographic emulsion.

Two important extensions of William Herschel's work on infrared radiation from the Sun were made by his son John. While working in South Africa in 1837–8 surveying the Southern Hemisphere for star clusters and nebulae, John Herschel carried out a very simple experiment to measure the total energy from the Sun falling on the Earth. The experiment consisted of a small waterbath in a blackened enclosure, with a thermometer to measure the temperature rise when sunlight was allowed to fall on it. After correcting for the absorbing effects of the Earth's atmosphere, the younger Herschel estimated that the total rate of flow of solar energy at the top of the atmosphere was about 1.4 kilowatts per square meter.[10] Two years later, John Herschel discovered the broad infrared absorption bands in the spectrum of the Sun caused by molecular absorption in the Earth's atmosphere, now called the *telluric* bands.[11] He used a sheet of paper blackened by soot, which was then soaked in alcohol. The dispersed spectrum of the Sun dried out the alcohol in the regions of the spectrum where the solar light is strong and left damp the regions of the infrared spectrum where four deep absorption troughs are found.[12]

William Herschel's discovery of infrared radiation was to prove of profound importance for astronomy, though it took well over a century to reach fruition. Today astronomers use the whole electromagnetic spectrum from radio to gamma rays. This discovery alone makes

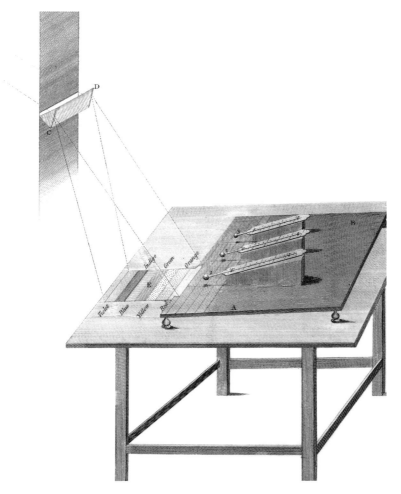

Figure 2.3. Herschel's experiment demonstrating the existence of infrared radiation from the Sun.

Herschel one of the giants in the history of science, even though it was not what he was celebrated for in his day.

Herschel's discovery of Uranus, his mapping of the Milky Way Galaxy and demonstration that it was probably just one 'island universe' among millions, his discovery of the huge depths of space and his vision of an evolving cosmos with new stars and galaxies being formed and eventually fading had a tremendous impact on the age he lived in. This has been brilliantly depicted by Richard Holmes in *The Age of Wonder*.[13] The stream of visitors to see Herschel's telescopes and meet the great man included Haydn and Byron, and each was influenced by him.

Joseph Haydn visited Herschel at Slough in 1791 and later claimed that this visit had helped him compose his great choral work *The Creation*. Byron wrote, after visiting Herschel in 1811 and looking through one of his telescopes: 'The Night is also a religious concern; and even more so, when I viewd the Moon and Stars through Herschel's telescope, and saw that they were worlds.' He is presumably referring to Herschel in 'Don Juan' when he writes:

> And though so much inferior, as I know,
> To those who, by dint of glass and vapour,
> Discover stars and sail in the wind's eye.
> I wish to do as much by poesy.

In perhaps his greatest poem, 'Darkness', written in 1816, Byron encapsulates the projected death of the Sun that Herschel's vision of an evolving cosmos implied:

> I had a dream, which was not all a dream.
> The bright sun was extinguished, and the stars
> Did wander darkling in the eternal space,
> Rayless, and pathless, and the icy earth
> Swung blind and blackening in the moonless air…

The most striking tribute of all to Herschel from the Romantic poets was Keats's 'On Reading Chapman's Homer':

> Then felt I like some watcher of the skies
> When a new planet swims into his ken;
> Or like stout Cortez when with eagle eyes
> He stared at the Pacific – and all his men
> Looked at each other with a wild surmise –
> Silent upon a peak in Darien.

On Herschel's death in 1822, his obituary said: 'As an Astronomer he was surpassed by none of the present age, and the depth of his research, and extent of his observations, rendered him perhaps second only to the immortal Newton'. His opening up of the invisible wavelengths for astronomy makes him one of the greatest astronomers who ever lived. But, as we shall see, progress in developing infrared astronomy was extremely slow. From Herschel's detection of infrared radiation from the Sun in 1800 to the reliable detection of even the very brightest stars in the infrared would take just over a century.

3

1800–1950: Slow Progress – the Moon, Planets, Bright Stars and the Discovery of Interstellar Dust

One of the reasons Herschel and his discovery of infrared light do not resonate more strongly in the history of science is that it took so long for infrared astronomy to develop. During the 150 years after Herschel, from 1800 to 1950, progress was extremely slow. There were two main reasons for this slow progress. Firstly, the night is not 'dark' at infrared wavelengths; in fact it is very bright. The Earth's atmosphere radiates strongly in the infrared as part of the greenhouse process, so even if we had infrared eyes we would have great difficulty picking out the stars against this bright foreground, even at night. The second and main problem was the slow progress in developing infrared detectors. Herschel's thermometers could detect infrared radiation from the Sun, but to detect anything else something better was needed. In this chapter I describe the slow progress in detector technology through the nineteenth and early twentieth centuries, the detection of infrared radiation from the Moon during the classic expedition of Piazzi Smyth to Tenerife, the efforts between 1870 and 1914 to detect radiation from bright stars, and the crucial discovery in the 1930s of extinction of visible light by interstellar dust.

The first step on the road to improved infrared detectors was made by a German physicist, Thomas Seebeck (1770–1831). In 1821, he made a discovery that led to the invention of the *thermocouple*.[1] He found that when a metallic strip is constructed of two different metals and then heated, a small electric current is generated in the strip. The thermocouple applies this discovery to the detection of heat, or infrared radiation, by measuring the current generated from a bimetallic strip when heated. It was this device, far more sensitive than the thermometer used by Herschel, that Piazzi Smyth would use to make the next major discovery in infrared astronomy, the detection of infrared radiation from the Moon, the second brightest astronomical object in the sky.

Charles Piazzi Smyth (1819–1900) (Figure 3.1) was born in Naples, where his father, a British naval captain, was serving. He was given the name Piazzi after his godfather, the Italian astronomer Giuseppe Piazzi, a friend of his father. His father subsequently settled in Bedford, England, and set up an observatory there, in which Piazzi Smyth first learned about astronomy. At sixteen he became an assistant at the Cape of Good Hope Observatory in South Africa, and there he observed Halley's comet and the Great Comet of 1843, and helped to confirm Nicolas Louis de Lacaille's measurement of the distance corresponding to one degree of longitude, which determines the circumference of the Earth. In 1845 he was appointed astronomer royal for Scotland, based at the Carlton Hill Observatory in Edinburgh.[2]

In 1856, Piazzi Smyth combined his honeymoon with an astronomical expedition to the island of Tenerife in the Canary Islands. The main purpose of the expedition was to test the quality of the peak of Guajara, close to the dominant volcanic peak of Mount Teide, for astronomical observations, using a 7-inch (18.4 centimetre) telescope provided by friends, and to verify Newton's speculation that 'a most serene and quiet air … may perhaps be found on the tops of the highest mountains above the grosser clouds'. Piazzi Smyth wrote:

> In the month of May, 1856, H.M. Lords Commissioners of the Admiralty, advised by the Astronomer Royal, were pleased to entrust me with a scientific mission to the Peak of Tenerife…. No sooner was the authorization known, than numerous valuable instruments were kindly proffered by many friends of astronomy; and one of these gentlemen, Robert Stephenson, M.P., immediately offered me the use of his yacht "Titania"; and by this, greatly ensured the prosperity of the undertaking.

> The object mainly proposed, was, to ascertain how far astronomical observation can be improved, by eliminating the lower third of the atmosphere. For the accomplishment of this purpose, an equatorial telescope and other apparatus were conveyed in the yacht to Tenerife, in June and July 1856. There – with the approval of the Spanish authorities (always ready in that island to favour the pursuits of scientific men of any and every country[3]) the instruments were carried up the volcanic flanks of the mountain, to vertical heights of 8900, and 10,700 feet, and were observed with during two months.[4]

Figure 3.1. Charles Piazzi Smyth.

Piazzi Smyth's book reporting on this expedition, *Teneriffe, an Astronomer's Experiment*, is notable for being the first to be illustrated by stereographic photographs. There is still an observatory today near the site used by Piazzi Smyth, and he can be said to have been the first to found a high-altitude observatory and to have pioneered the modern practise of placing telescopes at high altitudes for better observing conditions.

While on Guajara in 1856, Piazzi Smyth used a thermocouple to demonstrate infrared radiation from the full Moon. He gives a vivid account in his book:

> As the moon gradually rose higher and higher, some observations of its optical spectrum were made; chiefly remarkable for the quantity of blood-red light thereby proved to exist, in its innocent-looking blue, or rather violet-coloured rays. At last, about 11 o'clock P.M., the heat experiments were commenced. The moon was in unfortunately low declination, so as to have a meridian altitude of only 42°; but all other circumstances were eminently favourable. The air was pure, and perfectly calm, every one but myself had long gone to bed, the fires had been put out four hours previously, and their sites were a long way off, with stone walls between. So the apparatus being mounted on a small

pier inside the telescope enclosure, with a range of nearly 20 feet clear in front of it, with no artificial lights about (for I had found it possible to read off the graduation, and write down the figures by moonlight); and with no other sources of heat in the neighbourhood, than my own body, and that swathed abundantly in non-conducting flannel, and kept well away from the thermotic pile – I tried a preliminary experiment.

Holding my naked hand in front of the peculiar voltaic arrangement, and at a distance of three feet, there was an instant move of the magnetic-needle through 7°; and then bringing it within three inches, there was so large a deviation, that I had to wait a long time, before the needle had recovered from the disturbance.

How calm, how perfectly dead calm was the air all that time; not a breath could be felt; not a sound could be heard; there was the silence, and stillness of death. This degree of silence felt inappropriate on a high mountain; for on such, there is in general, so ceaseless a murmur of hundreds of torrents far and near, working their way downwards continually; and never for a second leaving off their bubbling, splashing, struggling onward – when powerful, even urging on the stones in their beds with a perpetual low, grinding, rumbling noise. But on Tenerife there was nothing of the sort; the absolute aridity of air and ground, had denied the existence of a single stream. A faint tinkle tinkle now and then, from a stray goat, was the only sound to be heard during this anxious period; and though the creature was far off, one could distinguish whenever it stopped to browse on some solitary retama bush, and then when it trotted off to find another.

The needle came at length to rest; so, quickly turning the voltaic cone, which had been directed to clear sky, on the moon – I anxiously watched the result – the needle scarcely moved. Lunar radiation was small indeed then, and I girded up my loins to try special means of observing, suitable to such a case. The plan decided on at last, was to take a large number of readings, at stated short intervals, combined with variations in the direction of the cone. Having at length obtained about two hundred such, in the course of an hour and a half, I was extremely well pleased to find, that the mean of the numbers indicated an undoubted heat effect, of about a third of a degree.

Had the recording instrument been a Fahrenheit's thermometer, the whole operation would have been concluded. But as a thermomultiplier's degrees may be almost anything, I immediately placed a candle on a stool, 15 feet in front of the pile; and observing it on exactly the same principle as the moon – there was given a heat effect of nearly one degree.

With this result, there need be no wonder at the failure of former observers in England, near the level of the sea, and before the day of the thermomultiplier – to obtain any instrumental evidence of the moon's radiation; for here, on so peculiarly favourable a position, with the luminary shining away quite blindingly, the heat was only one-third that of an ordinary candle, at the distance stated above.

This wonderful account brings to life Piazzi Smyth's night on the bare mountain, at the frontier of a new science. He captures the emotion, the excitement of the scientific experiment, and the dramatic landscapes in which astronomers work.

Piazzi Smyth also made observations at different altitudes and showed that the highest altitudes were best, thereby demonstrating that the Earth's atmosphere absorbs infrared radiation. He also studied the zodiacal light, the faint light seen at dawn and dusk along the zodiac caused by dust particles in the Solar System, and concluded from night-long observations that proposals that the zodiacal dust cloud encircled the Earth rather than the Sun were mistaken. *Teneriffe, an Astronomer's Experiment* is one of the very early popular books on science, as well as being a lively account of nineteenth-century travel in an unfamiliar landscape.

Lawrence Parsons (1840–1908), the fourth Earl of Rosse, son of the builder of the giant 72-inch 'Leviathan' telescope, followed up Piazzi Smyth's discovery of infrared radiation from the Moon. In 1870, he studied the Moon through its phases and estimated the range of temperatures of its surface through these phases. He also showed that it did not have an internal source of energy: 'The greater part of heat received from the Moon consists of solar heat that has been first absorbed by the lunar crust, and then given off in dark radiation. No evidence of cosmical heat was obtained.'[5]

One of the key tests Parsons did was to insert a sheet of glass into the beam of the telescope to demonstrate from the drop in signal that it really was infrared radiation falling on his detector. Glass is a strong absorber of infrared radiation, and this is the principle of the glass greenhouse. Optical energy from sunlight shines through the glass, but little of the infrared radiation from the contents of the greenhouse can escape and the temperature inside the greenhouse rises.

INFRARED RADIATION FROM STARS

After the Sun and Moon, it was natural for astronomers to try to detect infrared radiation from the brighter stars. In 1869, William Huggins (1824–1910), who was the first to study the optical spectra of stars, made the first attempt to observe stars in the infrared waveband, using an array of thermocouples called a *thermopile*. He used a technique that has come to be a staple of modern infrared astronomy and is now called 'chopping', in which observations of the desired target are alternated with observations of a nearby blank field. The emission from the atmosphere should be about the same in the two adjacent fields, so the difference should be just the radiation from the star. Huggins reported:

> The needle was then watched during five minutes or longer; almost always the needle began to move as soon as the image of the star fell on it. The telescope was then moved, so as to direct it again to the sky near the stars. Generally, in one or two minutes, the needle began to return to its original position. In a similar manner twelve to twenty observations of the same star were made.[6]

Huggins claimed to have detected the bright stars Regulus, Arcturus, Sirius and Pollux in this way. However, there is considerable doubt about whether he was really detecting infrared radiation with his thermopile, because he was using a refracting telescope, with a glass lens, and this would not have let much infrared radiation through.[7]

An important development for infrared astronomy was the invention of the *bolometer* in 1881 by Samuel Langley (1834–1906). Langley's bolometer used the fact that electrical resistance of a strip of the metal platinum is highly sensitive to temperature. The changes in resistance could be measured with great precision with a simple electrical circuit, and temperature changes as small as one-ten-thousandth of a degree could be recorded. Working from Mount Whitney in the Sierra Nevada range in California, Langley extended measurements of the Sun's spectrum from the 0.7–0.9 micron range studied by Herschel out to the much longer infrared wavelength of 5 microns.[8] With his sensitive detector, he was able to record over 700 distinct infrared *absorption lines* in the Sun's spectrum.[9]

Emission or absorption lines across a spectrum correspond to the characteristic wavelengths of particular atoms, and the strengths of the lines allow us to measure the *abundance* of different elements on the Sun or in a star, meaning how much there is of each particular element. Atoms, the basic building blocks of matter, have a dense nucleus composed of neutrons (which have no electric charge) and positively charged protons, surrounded by a cloud of light, negatively charged electrons. The atomic number of an element is the number of protons (which is also the number of electrons): 1 for hydrogen, 2 for helium, 6 for carbon, and so on. According to quantum theory, the electrons around a nucleus can only occupy discrete energy levels. If an electron makes a transition from a higher energy level to a lower level, the difference in energy is radiated away as a photon of light with a specific and recognizable energy (and hence with a specific wavelength). This gives rise to an *emission line*, a bright line across the spectrum of hot gas or a star, where some of the electrons have been excited to higher energy levels. The element sodium has a very strong emission line in the orange part of the spectrum, and this is what gives rise to the orange colour of neon street lamps. Similarly, if an atom of cool material between us and a star absorbs this specific energy, an electron can make the transition from the lower energy level to the higher one, resulting in an *absorption line*, a dark line across the spectrum.

Langley's bolometer, based on the temperature-sensitive electrical resistance of a variety of metals, alloys and crystals, has been in use for infrared astronomy ever since, for example in the far-infrared camera on the *Herschel Space Observatory*, launched in 2009. Towards the end of the nineteenth century, Ernest Nichols (1869–1924) developed a new kind of infrared detector, consisting of an evacuated glass bulb containing vanes suspended on a quartz wire, which he called a *radiometer*. It worked through the pressure of radiation acting on the vanes and making them rotate. Nichols used his radiometer to make the first reliable detection of infrared radiation from a star in 1901, just over 100 years after Herschel's first detection of infrared radiation from the Sun.[10] Nichols's radiometer was a big improvement on previous infrared detectors, and he detected small signals from the bright stars Arcturus and Vega. This radiometer was also used in 1924 by Charles Abbott (1872–1973) to carry out observations of nine bright stars at infrared wavelengths.[11] He took these observations a stage further by fitting his measured spectra with Planck's theoretical blackbody curves (Figure 1.5) so that he could deduce the surface temperature of the stars. Because he knew the distances of these stars, he could now use the known power output per unit area of a Planck blackbody to estimate their radii. While the radius of the bright star Procyon turned out to be similar to that of the Sun, the radii of the stars Rigel and Alpha Herculis were several hundred times larger, so they became known as *giant* stars. This was an important step forward in our understanding of stars.

Yet another detection device, the *vacuum thermocouple*, was optimized for astronomy by William Coblentz (1873–1962), who used it to study infrared radiation from stars[12] and the planets.[13] By 1914 he had observed more than a hundred stars, and George Rieke suggests that we should regard him as the first stellar infrared astronomer.[14] The most systematic infrared studies using the vacuum thermocouple were carried out by Edison Pettit (1889–1962) and Seth Nicholson (1891–1963) at the Mount Wilson Observatory in Los Angeles during the 1920s and 1930s. Altogether they observed 124 bright stars and greatly extended the results of Abbott and Coblentz.[15] They gave the first reasonably accurate estimates of the temperatures of stars and also made infrared observations of the planet Mercury[16] and carried out a systematic infrared study of the Moon through its phases,[17] demonstrating that the Moon's surface cooled down faster than expected for a solid surface. In 1948, Adriann Wesselink (1909–1995) carried out detailed theoretical modelling of these results and concluded that the Moon is covered by a layer of dust,[18] a deduction that was vindicated by the *Apollo* Moon landings 20 years later.

A major breakthrough in the technology of infrared detection was the development of the *lead sulphide (PbS) cell*, first by German scientists during World War II and then, following capture of the device, by the British Admiralty after the war. It is a type of bolometer, and infrared light falling on this crystalline material results in a small change in electrical resistance of the cell. The lead sulphide cell was sensitive to radiation at wavelengths from 1 to 4 microns and was a thousand times more sensitive than a thermopile. The Dutch American astronomer Gerard Kuiper (1905–1973) and his colleagues used this device in 1947 to carry out the first infrared spectroscopy of planets and bright stars.[19] Kuiper went on to found the Lunar and Planetary Laboratory at the University of Arizona, which was to become a major force in the emergence of modern infrared astronomy, and he also built a 61-inch infrared telescope at Mount Bigelow, Arizona, in 1965, which would be a dominant force in infrared astronomy for a decade.

In the United Kingdom, Peter Fellgett (1922–2008) used a lead sulphide cell and vacuum-tube electronics but achieved only modest improvements in sensitivity over Pettit and Nicholson. George Rieke points out that the weakness of Fellgett's work was his failure to use a differencing technique to subtract out the foreground emission from the atmosphere.[20] This was a missed opportunity for U.K. infrared astronomy, with Fellgett concluding in 1951 that 'there is little chance of detecting, in the infrared, stars that are too cool to be visible'.[21] This was to be proved massively wrong by Gerry Neugebauer and Bob Leighton's Two Micron Survey in 1969 (see Chapter 4).

INTERSTELLAR DUST

In 1930, the Swiss American astronomer Robert Trumpler (1886–1956) made a discovery that was crucial to the development of infrared astronomy, though in fact he made this discovery in the optical waveband. Trumpler was cataloguing star clusters with a view to estimating the size of the Milky Way. An important debate had taken place in 1920 on the nature of the spiral nebulae first identified by Lord Rosse with his 'Leviathan' 72-inch telescope in the 1840s. This debate hinged on the size of the Milky Way Galaxy. Harvard College Observatory astronomer Harlow Shapley argued that the spiral nebulae were part of the Galaxy, since firstly their distances were no greater, he thought, than his estimated size of the Galaxy: 300,000 light years. Shapley also pointed out that the distribution of the spiral nebulae was in some way connected with the Milky Way, since there were more of them towards the pole of the Galaxy than near its plane. Heber Curtis of Lick Observatory, on the other hand, argued like William Herschel that the spiral nebulae were distant galaxies of stars similar to our own Milky Way Galaxy.

Trumpler had examined three hundred clusters of stars and estimated their distances and sizes. To his surprise he found that the more distant clusters appeared to be about twice the size of nearby ones. Reasoning that this was implausible, he realized this could result from an intervening absorbing medium, which would make the distant clusters appear fainter and hence be assigned excessive distances and linear sizes. Trumpler identified the absorbing medium as *interstellar dust*, small solid grains of dust spread between the stars, and he went on to quantify the absorbing effect of this dust.[22] There had been earlier indications of gas between the stars through the presence of absorption lines of calcium and sodium in the spectra of some stars. In the plane of the Milky Way, the brightness of a star is diminished by the dust absorption by a factor of 1.2 for every thousand light years that the starlight travels towards the Earth. We call this dimming in the brightness of stars caused by absorption by interstellar dust *interstellar extinction*. Shapley was forced to revise his estimate of the size of the Milky Way downwards by a factor of three. The dust also explained why there appeared to be fewer spiral nebulae near the Milky Way itself since the absorption increases as we look from the pole of the Galaxy towards the plane, with a peak absorption towards its central plane.

Trumpler also found that stars of a given *spectral type*, and hence of a given surface temperature, looked increasingly redder with distance. At the end of the nineteenth century, astronomers working at the Harvard Observatory had classified the spectra of stars into different types according to whether they showed absorption or emission lines. They gave each type a letter from A to O on the basis of the complexity of the spectra. It was later realized that these types gave a sequence of surface temperatures, though not exactly in alphabetical order. The hottest stars were of types O, B, and A, the coolest of types K and M. The order of the spectral types OBAFGKM is remembered by the mnemonic 'Oh Be a Fine Girl/Guy Kiss Me'. Trumpler's discovery that stars of a particular spectral type looked redder at greater distances showed that the absorption of the intervening material was greater at blue wavelengths than it was at red wavelengths. This fits in well with the idea that the extinction and reddening of starlight are caused by small grains of interstellar dust. Between 1930 and 1948, Joel Stebbins (1878–1966), a pioneer of accurate measurements of the brightnesses of stars, studied interstellar reddening using hot and luminous stars of spectral types O and B, and this work was refined in collaboration with Albert Whitford (1905–2002). Their *interstellar extinction law* gave the dependence of reddening and extinction by interstellar dust on wavelength, with the extinction being roughly inversely proportional to wavelength from 0.36 to 1 micron.[23] So between the ultraviolet at 0.36 microns and the near infrared at 1 micron, the extinction drops by a factor of 3. As we move towards longer infrared wavelengths, dust clouds gradually become more transparent, and this explains the power of infrared and submillimetre astronomy in probing into the densest clouds of gas and dust. The interstellar extinction law established by Stebbins and Whitford remains valid to the present day, though it has been subsequently much refined and extended.

In 1963, the Dutch American astronomer Mayo Greenberg (1922–2001) reviewed what was known about interstellar dust.[24] He saw dust grains as playing several important roles: a negative one of extinction of starlight, and positive ones as tracers of physical conditions in the interstellar gas and through their physical interactions with the other components of the interstellar gas. In discussing the origin of dust grains, Greenberg mentions that at first the idea had been that they had originated like the meteors as fragments of larger bodies and were perhaps metallic. However, this was before it was realized that meteors are probably not of interstellar origin but are localized to the Solar System. The Dutch astronomer Bertil Lindblad (1895–1965) had proposed in 1935 that grains grow by condensation in interstellar gas.[25] Today we think that most dust grains originate in outflows from dying stars and then grow icy mantles within dense clouds of interstellar gas.

The nature of interstellar grains remained mysterious, with rival suggestions of large molecular structures[26] and graphite particles.[27] Greenberg himself preferred the idea of 'dirty ice', water ice with contamination by compounds of the other main atomic species in the interstellar gas. He correctly recognized that dust grains would be at different temperatures, depending on their radius and on the brightness of the radiation falling on them.

Trumpler's discovery of interstellar dust was complemented in 1951 by the detection of interstellar hydrogen gas through the detection of the 21-centimetre radio spectral line of atomic hydrogen by Harold Ewen and Edward Purcell at Harvard University. The 21-centimetre spectral line had been predicted by Hendrik van de Hulst in 1944, and its detection opened up the way to map the gas between the stars using radio telecopes. The spiral structure of the Galaxy could be traced and the rotation of the Galaxy mapped out. Astronomers refer to the gas and dust between the stars as the *interstellar medium*.

THE STRUCTURE AND EVOLUTION OF STARS

Infrared observations had helped to tie down the surface temperatures of stars, and in some cases their radii. This was a crucial step in the classification of stars and in the gradual understanding of their internal structure and evolution. The central figure in the development of the theory of the structure and evolution of the stars was Arthur Eddington (1882–1944). His ideas were brought together in his 1925 book *The Internal Constitution of the Stars*.[28] Eddington showed that the central temperatures of stars had to be millions of degrees and that the flow of radiation through a star is dictated by the composition and opacity, the capacity to resist the flow of radiation, of the stellar material. He also made the key suggestion that the likely energy source for stars, what kept them shining, was nuclear processes. During the 1930s the first steps to understanding these nuclear processes were made, and it was realized that the dominant constituent of stars was hydrogen, with helium as the next most abundant element. The most important nuclear process in stars was the nuclear fusion of four atoms of hydrogen to make an atom of helium. Because an atom of helium is slightly lighter than four atoms of hydrogen, the energy associated with that mass difference, through Einstein's famous equation $E = mc^2$, is released in the process, and this is what keeps the Sun shining.

It was not until 1946 that Fred Hoyle (1915–2001) made the first realistic assessment of the fraction of elements heavier than helium in stars, of order only 1%. Astronomers talk about atoms heavier than carbon as the *heavy elements*.[29] During the 1940s and 1950s understanding of the structure and evolution of stars of different masses grew, especially in the difficult modelling of the late stages of massive stars leading up to their final explosions as supernovae. In 1957, two independent studies drew together a picture for the origin of almost all the chemical elements in the interiors of stars and in explosive nucleosynthesis during supernova explosions.[30] Nuclear processes in stars could account for the observed abundances of all the elements except helium and other light elements, which were subsequently shown to be made either during the hot phase of the Big Bang (see Chapter 6) or through the impact of energetic cosmic rays on helium nuclei in the interstellar gas.

By the late 1950s astronomers had achieved a good understanding of the structure and evolution of stars and their role in generating almost all the chemical elements. After 150 years of development since William Herschel's discovery, infrared astronomers were beginning to play a part in the understanding of stars and the material spread through the interstellar space between them. Now they were poised to open up entirely new areas of research, especially in studies of dying stars and stars in the process of being born.

4

Dying Stars Shrouded in Dust and Stars Being Born: The Emergence of Infrared Astronomy in the 1960s and 1970s

PIONEERS WORKING FROM MOUNTAINTOPS

In this chapter, I describe how pioneers working from mountaintops discovered new types of infrared stars, which turned out to be either dying stars shrouded in dust or stars in the process of being born. The leaders of this infrared revolution were the Americans Frank Low (1933–2009) and Gerry Neugebauer (b.1932). First efforts to map the sky with telescopes on balloons and rockets began the push to observe at longer wavelengths inaccessible from the ground. And theorists began to use the observed distributions of brightness with wavelength in the new infrared sources to model the distribution and properties of the interstellar dust in clouds around stars and spread through interstellar space.

The first pioneer of the modern age of infrared astronomy was Harold Johnson (1921–1980) (Figure 4.1). He and his group, then at the University of Texas at Austin, began in the late 1950s to work with lead sulphide detectors, which they cooled with liquid nitrogen to 77 K (−196 °C). Cooling made the detectors much more sensitive, and over the next few years Johnson's group observed several thousand stars. Johnson had earlier defined the standard optical observing bands, which are denoted U (ultraviolet), B (blue), V (visual), R (red) and I (infrared). He now extended these bands further into the infrared, defining the J (1.25 microns), K (2.2 microns), L (3.6 microns) and M (5.0 microns) bands,[1] which fell in ranges of wavelengths, or windows, where a high fraction of light is transmitted through the Earth's

---

I have made extensive use of the historical review by Low, Rieke and Gehrz (2007) in this chapter. There are also valuable reviews by Greenberg (1963), Johnson (1966b), Neugebauer, Becklin, and Hyland (1971), Rieke and Lebofsky (1979), Savage and Mathis (1979), Mathis (1990), Draine (2003) and Price (2009).

Figure 4.1. Harold Johnson.

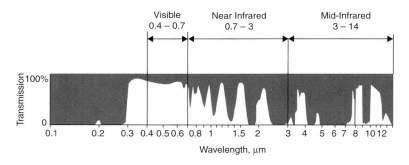

Figure 4.2. Atmospheric transmission across the ultraviolet, visible and infrared. The white areas show the wavelength windows that can be used for astronomy.

atmosphere (Figure 4.2), especially as viewed from mountaintops like those of the Catalina Mountains of Arizona, where Johnson and his group worked.[2]

In 1960, Frank Low (Figure 4.3), a young physicist working on phenomena at very low temperatures at Texas Instruments Research

Figure 4.3. Frank Low.

Lab in Dallas, Texas, invented the *gallium-doped germanium bolometer*, which operated at liquid helium temperatures, close to absolute zero.[3] The detector is a crystal of germanium into which a small amount of gallium (the 'doping') has been introduced. When cooled to very low temperatures, its electrical resistance is very sensitive to temperature and will change depending on how much infrared radiation it absorbs. This detector permitted observations at much longer wavelengths, all the way from 1 micron to 1 millimetre. Low's work attracted a lot of interest from astronomers, and several visited him in Dallas. One of them was Carl Sagan, who wanted him to build a bolometer system so NASA could fly an infrared spectrometer on a balloon to look for organic molecules on Mars, which might indicate signs of life there. Although none were found, the project had the important consequence that Low needed to improve on the fragile pyrex dewars, or vacuum flasks, used for storing the liquid helium. He designed a metal dewar which is now known as the Low dewar and has been used by infrared astronomers ever since. Soon afterwards Low met Harold Johnson, and they teamed up to define a new infrared observing band, N (10 microns).[4] Soon after that a further band at 20 microns, Q, was defined. The fraction of infrared radiation which gets through the Earth's atmosphere in these two wavelength windows is not great at sea level, but they can be used at dry, high-altitude sites.

Frank Low has recalled his first efforts at 10-micron observations using the 82-inch telescope at McDonald Observatory in Texas:

> As I recall, the dome at McDonald Observatory could be reached only by foot on rather narrow and steep steps. One day as Arnold [Davidson] and I were leaving the dome, we crossed paths with a distinguished looking optical astronomer who was surprised to find two young persons there. He wanted to know who we were and why we were there. We identified ourselves and told him we were working with Harold Johnson in the infrared. His response was that he could not understand why we were spending our time waiting for clear weather when we were not going to get any results in the infrared.[5]

Eventually the clouds cleared and in one clear night Low made the first, pioneering 10-micron observations of 24 stars and several planets.[6] Johnson's group made observations in all the photometric bands from U to Q (0.36–20 microns). Around 1963, they switched from using lead sulphide cells to the more sensitive indium antimonide detectors for work from 1 to 5 microns.

In 1966, Johnson collected together colours in the ten wavelength bands from the ultraviolet to the infrared for almost one thousand stars of different types, and for each type he calculated the best-fitting surface temperature.[7] This was a fundamental advance in accurate knowledge of the properties of stars of different types. By comparing normal stars with those that are reddened by interstellar dust, Johnson was able to extend Whitford's interstellar reddening law far into the infrared, showing that dust clouds gradually become transparent as we look to longer wavelengths.[8] This showed the great promise of infrared astronomy in being able to peer through clouds of dust that might be completely opaque at optical wavelengths. In fact some of these stars were reddened not by interstellar dust but by clouds of dust surrounding the stars, which became known as *circumstellar dust shells*. These stars appeared to have peculiar reddening laws, and at first this was attributed to absorption by dust with unusual properties. Gradually it became clear that the anomalous behaviour resulted from infrared light being radiated from dust clouds surrounding the stars.[9]

THE TWO MICRON SURVEY

In the mid-1960s, two astronomers from the California Institute of Technology, 'Caltech', started on an ambitious campaign to survey the whole northern sky in the infrared at 2.2 microns. Gerry Neugebauer (Figure 4.4) and Bob Leighton built a 62-inch parabolic mirror by filling an

Figure 4.4. Gerry Neugebauer.

aluminium bowl with epoxy resin and then rotating it at constant speed until the epoxy set into the desired parabolic shape. This was installed in a simple telescope on Mount Wilson, California, only an hour's drive from Caltech, and the survey was carried out over several years.[10]

The output from their cooled lead sulphide detector was recorded automatically on long rolls of paper, and these 'strip charts' were analyzed with the help of a part-time army of volunteers. Neugebauer only wanted to include extremely reliable sources in the Caltech catalogue, which became known as the IRC catalogue (short for InfraRed Catalog).[11] Sceptical colleagues had assured Neugebauer and Leighton that they were wasting their time and would see only normal stars, but there was a wealth of interesting and unusual objects in the Two Micron Survey, and these began to put infrared astronomy on the map.

Prior to publication of the main catalogue, Neugebauer and Leighton, with D.E. Martz, announced two of the more extreme sources, bright in the infrared but faint in the optical.[12] These became known as NML Taurus and NML Cygnus, after the initials of the authors and the constellations where the objects were found. Both turned out to be red giant stars surrounded by dense clouds of dust.[13] Altogether there were about 50 infrared sources in the survey with colours corresponding to temperatures of 1000 K or less, compared with the surface temperatures of stars, which range from 2000 K for the coolest red giants to 40,000 K for the hottest blue supergiants. Although most of the sources in the survey (93%) could be found in a well-known catalogue of faint red stars,[14] what made the survey really interesting were those stars that showed excess infrared radiation relative to the normal thermal radiation from the surface of the stars and the handful of new infrared sources not in known stellar catalogues.[15] These unusual objects showed that infrared astronomy, up to then considered only an adjunct to optical astronomy, was now going to open up new areas of research.

The main phenomenon the new sources from the Two Micron Survey were revealing was mass loss from red giant stars. Dying stars were spewing out clouds of dust and gas in steady winds from their surfaces. A star spends most of its life fusing hydrogen to helium in its hot core. During this phase stars define a line in a plot of *luminosity*, or total power output from the star, against colour or surface temperature, called the *main sequence* (Figure 4.5). Gradually the abundance of hydrogen in the centre declines and that of helium grows. When stars of between one-half and eight times the mass of the Sun exhaust their prime nuclear fuel, hydrogen, in their core, they start to burn hydrogen in a shell around the helium core and as they do so they change their structure. The main process of heat transfer from the centre of the star to its surface changes from the direct flow of radiation through the gas to convective up-and-down motions of the gas, like a bubbling saucepan. The star grows dramatically in size, becoming cooler and redder at the surface. In a plot of luminosity against colour (the 'Hertzsprung–Russell' diagram; see Figure 4.5), they start to ascend the 'red giant branch', becoming cool, red, large and luminous. After a while the temperature at the centre becomes hot enough for helium burning to start, and after a brief convulsion the star continues a second ascent of the red giant branch.[16] The whole red giant phase lasts only a few million years, compared with the ten billion years that a star like the Sun spends burning hydrogen on the main sequence. The lifetime of a star depends very sensitively on its mass. The lifetime of a star of mass one-tenth of the Sun would be three thousand billion years, while for a star ten times the mass of the Sun it would be only 30 million years.

During this second red giant phase for stars similar to the Sun, the star becomes progressively more unstable dynamically and starts to pulsate in and out. As it does so, the brightness and surface temperature of the star vary over a period from 100 to 500 days. The star starts to lose material from its surface in a strong, steady, spherical wind. As this wind streams away from the star and cools, solid material condenses in the form of small dust grains, whose composition depends on what products from nuclear burning have been dredged from the centre of the star to the surface. These red giant stars are the factories making the dust grains of interstellar space.

At the end of the red giant branch phase, when the star is at its most luminous, it undergoes one final convulsion and ejects the whole of its outer layers to form the spectacular display of a *planetary nebula*. Of the bright nebulae that can be seen with binoculars, those in the famous list compiled by Charles Messier in the eighteenth century, planetary nebulae are some of the loveliest. They have nothing to do with planets, though. After ejection of the star's outer layers, the hot compressed core, exhausted of its nuclear fuels, becomes a compact, dead, white dwarf star and steadily cools off over billions of years. Many of the most interesting new infrared sources, those with the strongest infrared excess compared with their visible light, were red giant stars surrounded by dense shells of dust and gas, including some in the last stages of ejecting their surface layers. Several planetary nebulae were also prominent infrared sources and turned out to have very interesting spectra.

Stars more massive than about eight times the mass of the Sun have an even more dramatic death. While burning hydrogen they are very blue, luminous supergiant stars. After exhausting the hydrogen in their cores, they fuse helium to carbon, nitrogen and oxygen and then proceed to manufacture heavier elements such as silicon, magnesium and eventually iron, developing an onion structure of layers of different compositions and in which different nuclear fuels are being burned. However, when the star has developed an iron core, nuclear burning comes to an end since the fusion of iron absorbs energy rather than generating it. The core collapses to a very compact state to form a neutron star or, if the original star is more than about 20 times the mass of the Sun, a black hole, and the outer layers of the star are violently expelled in a supernova explosion. Some of the very brightest new infrared sources turned out to be very interesting supergiant stars, which may be on the brink of becoming supernovae. Two examples are VY Canis Majoris and Eta Carinae.

VY Canis Majoris is a red supergiant star, with a radius of around 2000 times that of the Sun. It is the largest known star and is one of the most luminous known, at four hundred thousand times the luminosity of the Sun. Its mass is 30–40 times that of the Sun. It was first

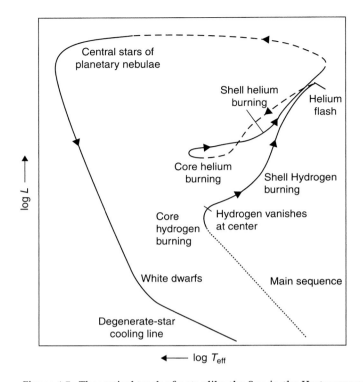

Figure 4.5. Theoretical track of a star like the Sun in the Hertzsprung–Russell diagram, a plot of the luminosity of the star versus its surface temperature (with hotter temperatures plotted to the left). The sloping, broken line indicates the main sequence, where stars burning hydrogen at their centres lie. As the Sun exhausts hydrogen at its centre, it moves to a higher luminosity and cooler surface temperature, beginning its ascent up the red giant branch. When the centre of the star becomes hot enough, helium ignites and the star moves rapidly to the left before resuming its ascent of the giant branch. Finally, the outer layers of the star are ejected in a planetary nebula event, leaving a degenerate white dwarf core, which gradually cools off.

catalogued in 1801 and is known to have been fading since 1850. In visible light it is surrounded by a very prominent cloud of hot gas and dust, which is the remnant of successive outbursts. It will explode as a supernova in the next 100,000 years.

Eta Carinae is an even more extreme supergiant star, a double-star system with a combined luminosity 4 million times that of the Sun and a combined mass around 120 solar masses (we use the shorthand 1 *solar mass* for a mass equal to that of the Sun), with the larger star believed to be 90 solar masses, close to the very highest mass that a single star can have. Eta Carinae was first catalogued by Edmund Halley in 1677 and underwent a giant eruption in 1841. It is surrounded by a dramatic double nebula, known as the Homunculus Nebula. Eta Carinae is expected to explode as a supernova within the next twenty thousand years, a rather short time in astronomical terms. With its nebula of hot gas and the surrounding dust cloud from which it formed, it is one of the brightest infrared sources in the southern sky. Both VY Canis Majoris and Eta Carinae will probably leave black hole remnants when they explode and die.

A supernova explosion within about one thousand light years could affect the Earth's atmosphere. Gamma rays from the explosion might result in depletion of the ozone layer, which shields us from the harmful effects of the Sun's ultraviolet radiation. However, at distances of 4500 and 7500 light years, respectively, VY Canis Majoris and Eta Carinae will have little effect.

THE MINNESOTA–SAN DIEGO AXIS

An important collaboration on infrared instrumentation, infrared observations and theory developed in the mid-1960s between Ed Ney and Nick Woolf at the University of Minnesota and Fred Gillett (1937–2001) and Wayne Stein at the University of California at San Diego.[17] Gillett and Stein collaborated with Frank Low to develop a 2.8–14-micron infrared spectrometer until it was destroyed in a freak accident. Fred Gillett had been observing at Mount Bigelow in Arizona and left the mountain to escape a blizzard, leaving the spectrometer in a trailer. When the weather cleared, the trailer was nowhere to be seen. The weight of the snow had flexed the roof of the trailer and snapped the gas heater supply line. The resulting fire had reduced the trailer and spectrometer to ashes, and the astronomers had to start again. Initially they used the spectrometer to study planets and stars,[18] but they knew that infrared spectrometry would also provide a key diagnostic to understanding the new infrared sources.

Ed Ney (Figure 4.6) built a dedicated infrared telescope, the O'Brien Observatory, in the high St. Croix hills in Minnesota. Though not a high-altitude site, it provided conditions dry enough for infrared astronomy in the winter. Later the Minnesota and San Diego groups built the Mount Lemmon Infrared Observatory in Arizona, which began operations in 1970. Fred Hoyle negotiated a $100,000 (U.S.) contribution from the United Kingdom in return for training aspiring U.K.

Figure 4.6. Ed Ney explaining cometary orbits at the O'Brien
Observatory. Bob Gehrz is on the right.

infrared astronomers. The Minnesota group was also heavily involved
in the development of the Wyoming Infrared Observatory, with Bob
Gehrz becoming director in 1972. The 92-inch (2.3 metre) Wyoming
Infrared Telescope began work in 1977.[19]

In 1969, Nick Woolf and Ed Ney made an important breakthrough
in understanding the dust surrounding stars. They drew attention to
what turned out to be a very significant broad emission feature in the
spectra of some infrared sources at a wavelength of around 10 microns
which could be interpreted as emission from dust composed of small
silicate grains,[20] material that would be similar in composition to that
of terrestrial rocks.

This was the first clue to the nature of the dust around stars.
Robert Gilman of the Goddard Institute for Space Studies in New York
calculated what solid materials would be likely to condense out of a
spherical wind of gas flowing away from the surface of a star and cooling down.[21] In a profound insight, he found that the outcome depended
crucially on the relative amounts of carbon and oxygen in the star's
atmosphere. If there was an excess of oxygen, then the carbon would

combine with oxygen to form carbon monoxide molecules and the rest of the oxygen would combine with iron, silicon, aluminium and magnesium to form iron, aluminium and magnesium silicates. If carbon predominated, then the main product would be graphite dust grains. While most red giant stars are oxygen-rich, a subclass of cool 'carbon' stars is known in which the carbon abundance is higher.[22] So the nature of the circumstellar dust clouds depended on whether the carbon ended up as carbon monoxide gas or in graphite dust grains. The silicate identification of the 10-micron feature was strengthened when Frank Low found in the spectrum of Betelgeuse (Alpha Orionis) a second emission feature, at 20 microns, which is also characteristic of silicates.[23] The silicate features turned out to be a very widespread phenomenon and were also found in absorption spectra of the interstellar dust and in the emission spectra of comets.

In 1971, Bob Gehrz and Nick Woolf showed that the wind of gas and dust being blown off red giant stars was driven by the pressure of radiation from the star pushing on the dust grains. They made a rough estimate of the total amount of dust being driven off such stars in the Galaxy and showed that it was enough to provide all the dust in the interstellar medium.[24]

STARS BEING BORN

So far the story has been about stars in the process of dying. Even more interesting discoveries were to emerge from infrared studies of regions where stars are being born. Prior to the emergence of infrared astronomy in the 1960s, two distinct types of young stars were known. T Tauri stars are very cool, low-luminosity stars found in dust clouds such as the nearby Taurus dust cloud. They are believed to be stars like the Sun in the final stages of contraction, close to the point where their core heats up sufficiently for nuclear fusion of hydrogen to start.[25]

The second type of young stars known are the hot, massive supergiants of spectral types O and B. These are ten to one hundred times more massive than the Sun and are thought of as young because they live such a short time in astronomical terms, only a few million years, so that they are in a sense always young. They are found in loose clusters of a few hundred stars, known as 'open' clusters (in contrast with the much more compact and dense 'globular' clusters, which contain millions of older stars), and in smaller groups of a few stars, called 'OB associations', consisting of just a few massive stars

Figure 4.7. Gerry Neugebauer, Eric Becklin and Gareth Wynn-Williams.

of spectral type O or B. The most famous of the open clusters is the Pleiades cluster, seven of whose stars are visible to the naked eye. And the most famous OB association, not very far away on the sky, is the Orion Trapezium association, a group of four stars embedded in the famous Orion Nebula.

The Orion Nebula is the central object in the Sword of Orion and looks slightly fuzzy to the naked eye. With binoculars or a small telescope, its extended nature becomes clear. It was first discovered by Nicolas-Claude Fabri de Peiresc in 1610 and sketched by Christiaan Huygens in 1656. Charles Messier recorded it as number 42 in his 1774 list of nebulous objects, so it is also known as Messier 42, or M42 for short. When William Huggins studied the spectrum of this nebula in 1865, he noticed emission lines characteristic of very hot gas, and this demonstrated for the first time the gaseous nature of the nebula. Robert Trumpler, in 1931, first noted that the gas is heated and 'ionized' by the four stars of the Trapezium cluster. An ionized gas is one in which the atoms have had their outer electrons stripped off by heating or ultraviolet radiation. The Orion Nebula is therefore a cloud of mainly hydrogen ionized by hot, massive stars, and we call this type of hot, ionized gas cloud an *H II region* (pronounced 'aitch-two').

In 1967, Eric Becklin and Gerry Neugebauer (Figure 4.7) mapped the Orion Nebula at infrared wavelengths. They found a bright infrared object close to the Trapezium, but with no counterpart visible on optical images.[26] It had a total luminosity about 1000 times that of the Sun. This object became known as the 'Becklin–Neugebauer (or BN) object' and had a tremendous impact because it opened up the idea that new stars, forming inside a dense cloud of dust and molecules, might be so heavily shrouded in dust that they were detectable only at infrared wavelengths.[27] Gradually the picture emerged of clusters of massive OB stars forming out of dense, opaque clouds of gas and dust, the bulk of their visible and ultraviolet light being absorbed by dust and emitted in the infrared. The massive stars ionize the gas around them to form an H II region, which grows in size, eventually breaking out of the original cloud of dust and molecules. At this stage we have something that looks like the Orion Nebula, an ionized region on the front surface of a dense cloud of gas and dust where new stars are forming.

The Orion cloud contains newly formed massive stars like the Trapezium cluster, massive stars in the process of formation like the BN object, and also low-mass stars in the process of formation like the T Tauri stars. On the other hand, the nearby star-forming cloud in Taurus where the T Tauri stars had originally been discovered contained no stars more massive than two solar masses.[28] From this contrast, the idea emerged that the formation of low-mass and high-mass stars is the result of two independent processes: low-mass star formation is a continuous process in all gas clouds, whereas the formation of high-mass stars requires clouds with much denser cores and also needs some kind of trigger to start the star formation.[29] In the more spectacular spiral galaxies, this trigger takes the form of a spiral compression wave rotating through the galaxy's gas. This is driven by tidal interaction with a companion galaxy, and results in the dramatic spiral patterns we see in photos of galaxies, which trace out regions where high-mass stars are forming.

So now we have seen the complete cycle of gas, dust and stars in the Galaxy, with stars forming out of dense clouds of gas and dust, evolving and then dying. As they die they eject huge amounts of their outer layers either in a dense wind or in a supernova explosion. When this ejected gas cools sufficiently, new dust grains condense out and the interstellar gas is replenished by this new gas and dust. Gradually the gas and dust assemble into clouds that become ever denser and the cycle starts again. An atom of carbon in our body, for example, has been through this whole cycle at least once, perhaps several times, so we really are made of stardust.

EARLY INFRARED OBSERVATIONS OF THE GALACTIC CENTRE

The year after their discovery of the BN object, Gerry Neugebauer and Eric Becklin decided to look for infrared radiation from the centre of our Milky Way Galaxy using the 62-inch telescope on Mount Wilson. Strong radio emission had been detected from a region of Sagittarius known as Sgr A in 1966. As Neugebauer writes: 'We had actually detected it with our 62-inch telescope several years earlier, but did not know it. I went to an optical astronomer, who is very famous, and asked him for the coordinates. He thought, "These guys don't know anything about it." He gave them to me but did not tell me that they weren't precessed,[30] so we missed it.'[31]

Eric Becklin, Neugebauer's graduate student, had been working with Neugebauer and others to develop a detector system that was more sensitive than the Two Micron Survey instrument and had a night on the 24-inch telescope to try it out. Neugebauer asked him what he was going to do with it. 'Look for the Galactic Centre,' said Becklin. 'Don't bother, it isn't there,' Neugebauer advised. However, it was. A signal at 2.2 microns was detected on the first scan. Becklin recalled, 'I thought maybe the Galactic Centre was just below the survey limit or there was a confusion of multiple sources. It never occurred to me that the position looked at on the survey was wrong.' The peak infrared position turned out to be very close to the bright source of radio emission Sgr A.[32] This radio source is now known to be a supermassive black hole (Plate IV) with a mass of several million solar masses right at the very centre of the Galaxy.[33] In Chapter 9 we will see how infrared studies of the Galactic centre demonstrated that this really is a black hole and measured its mass.

EXTRAGALACTIC SOURCES OF INFRARED RADIATION

The Two Micron Survey detected no objects outside our Milky Way Galaxy. It was simply not sensitive enough to reach the expected emission from starlight in even the brightest galaxies.[34] However, the first extragalactic object to be detected in the infrared was not a normal galaxy, and its infrared emission was certainly not caused by starlight.

In the 1950s and early 1960s, radio astronomers had discovered that some galaxies are exceptionally powerful emitters of radio emission. This was believed to result from a process called synchrotron radiation, in which electrons moving very close to the speed of light spiral in a large-scale magnetic field in the galaxy, emitting a tiny beam of radiation as they do so. The astronomers also found that some of the objects in their surveys looked stellar on optical images, and the term 'quasi-stellar radio source', or 'quasar', was coined. In 1963, Cyril Hazard identified the radio source 3C 273 as a relatively bright stellar object, but it was not a known star. Maarten Schmidt obtained a spectrum of the object at the Mount Palomar 200-inch telescope, and this spectrum showed bright emission lines. He puzzled over the wavelengths of the lines and realized they made sense if he assumed they were all shifted towards the red end of the spectrum by 15.8%. Now it had been known since Edwin Hubble's 1929 discovery that the universe was expanding (see Chapter 6) that the light from distant galaxies is redshifted because these galaxies are moving away from us as the universe expands. The larger the redshift, the farther away the galaxy is. If the redshift of 3C 273 was cosmological, caused by the expansion of the universe, then its distance was two thousand million light years and the object was one of the most distant in the universe known at the time. Because it was so far away, it had to be exceptionally luminous both at radio and at optical wavelengths, where it was more luminous than any known galaxy. Other quasars were soon found to have even higher redshifts and distances, and it was clear that a very interesting class of object had been discovered.

Within a year of the discovery of 3C 273, in 1964, Frank Low and Harold Johnson had detected it in the infrared and so, rather strangely, one of the most distant objects known in the universe at that time became the first extragalactic object to be detected at infrared wavelengths.[35]

The emission-line spectra of quasars linked them to a class of galaxies that had been discovered 20 years earlier by Karl Seyfert.[36] Both kinds of objects had strong emission lines in their spectra, often broadened, which showed that the gas responsible for the emission lines was in very rapid motion. Seyfert galaxies have a bright, compact optical nucleus, as if they harbour a miniature quasar. Soon several quasars and a number of Seyfert galaxies had been detected in the infrared.[37] For the Seyfert galaxies it was clear that the infrared radiation was much brighter than the starlight in the galaxies, though it was not understood what was causing this.

In 1972, George Rieke (Figure 4.8) and Frank Low announced the results of a major study in which they had observed 57 galaxies at a

Figure 4.8. George Rieke.

wavelength of 10 microns.[38] They defined a new category of 'ultra-high-luminosity' galaxies. For the most luminous of these, the Seyfert galaxy Markarian 231, they quoted an infrared luminosity of 230 billion solar luminosities, which they revised three years later, following measurements at 34 microns, to four trillion (4×10^{12}) solar luminosities.[39] This was still a bit of a guess since it had still not been observed beyond 34 microns, and it turned out that most of the luminosity of Markarian 231 is emitted at wavelengths longer than this. But Rieke and Low were justified in claiming that Mk 231 had an exceptionally high luminosity, of order one hundred times greater than our Milky Way Galaxy.

THE UNIDENTIFIED INFRARED BANDS

The unexpected discovery in 1973 of an emission feature at 11.3 microns from two planetary nebulae by Fred Gillett and his colleagues at the University of California at San Diego[40] opened an interesting chapter of infared astronomy. They had been systematically developing infrared spectroscopy at 2–14 microns and they naturally first observed bright stars.[41] The normal stars showed broadly blackbody spectra with dips in the infrared caused by molecular absorption. When Gillett and his team looked at the stars with excess infrared emission

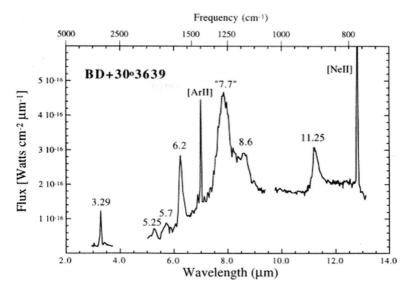

Figure 4.9. Unidentified infrared bands in the infrared spectrum of a planetary nebula. The bands, at 3.3, 6.2, 7.7, 8.6 and 11.25 microns, were first discovered by Gillett, Forrest and Merrill in 1973.

found in the Two Micron Survey, they had found the 10-micron emission feature associated with silicate dust. The planetary nebulae were a shock because they showed a series of new broad but strong features at 3.3, 6.2, 7.7, 8.6 and 11.25 microns (Figure 4.9). None of these had an obvious chemical explanation, and they became known as the *unidentified infrared bands*. The same features began to be found in a variety of situations where interstellar dust was strongly illuminated by ultraviolet light. As well as being found in planetary nebulae, the unidentified infrared bands were found in regions of hot ionized gas such as H ɪɪ regions and reflection nebulae, and in the diffuse gas between stars. It would be more than ten years before the explanation for these features was found.

THE U.S. AIR FORCE ROCKET SURVEYS

So far I have talked about infrared observations from mountaintops, and this restricts studies to the wavelength windows accessible from the ground. To make progress at longer wavelengths it was necessary to get above the Earth's atmosphere. In 1975, Russ Walker and Steve Price announced the results of a survey of the sky at 4–27 microns made from Aerobee rockets, carried out under the auspices of the U.S. Air Force.[42] Prior to this rocket survey there had been a dozen U.S. Army and Air Force rocket flights testing the performance of infrared sensors in space. The military interest in astronomical infrared sources was of course because the sources created a confusing background against which the military were trying to detect satellites and rocket launches.

Altogether Walker and Price used ten rocket flights, each allowing a few minutes of observation above the Earth's atmosphere. The rocket carried a telescope that had a mirror 16.5 centimetres across and was cooled with liquid helium. In the focal plane of the telescope there was an array of 24 high-performance infrared detectors. The AFCRL catalogue, based on the first seven rocket flights, listed over 3000 sources found all over the sky (AFCRL stood for Air Force Cambridge Research Laboratory). Many of these sources were associated with known stars. Several of the leading infrared astronomers of the day spent the next few years trying to observe the new, unidentified sources, but with a very poor success rate of only about one in five.[43] Some of the AFCRL sources were extended and so were not detected in ground-based chopped observations. Another problem was that some of the sources listed had only been detected once and were probably not real astronomical sources, so the catalogue had to be revised.[44]

Although this rocket survey had its teething problems, it was important as the first survey of the sky at wavelengths between 2 microns and 6 centimetres, in the radio band, a gap in wavelength of a factor of 30,000. It was the first hint of what an all-sky survey of the mid- and far-infrared sky had to offer. Price and Walker's survey had a major influence on research in infrared astronomy between 1976 and the launch of *IRAS* in 1983. Their survey influenced the design of the *IRAS* data processing and also the instrument design of the DIRBE instrument on the *Cosmic Background Explorer* (*COBE*, Chapter 8).

What kinds of sources had they detected? Most of the shorter-wavelength (4 micron) sources were stars, and in fact 83% could be associated with sources from the Two Micron Survey. However, these percentages dropped to 66%, 46% and 26% at 11, 20 and 27 microns, so at the longer wavelengths most of the sources were new. The new sources were strongly concentrated towards the Galactic plane and could often be associated with H II regions (clouds of ionized gas around newly formed, very massive stars) or with dense molecular clouds which had been identified from surveys for the carbon monoxide molecule (see Chapter 5). So these were all connected to the formation of new, massive stars. Other new sources were found to be dying stars: cool red giant stars undergoing strong mass loss and planetary nebulae. Some of the most interesting new sources were stars at the top of the red giant branch, surrounded by dense shells of gas and dust, on the very brink of ejecting their outer layers as planetary nebulae. A few sources could be identified as hot gas clouds within our neighbouring galaxy, the Large Magellanic Cloud, and the galaxies M31, M82 and NGC 253 were also detected. However, the real opening up of the extragalactic infrared universe had to wait for the *Infrared Astronomical Satellite* (*IRAS*) mission in 1983.

MODELS FOR INFRARED SOURCES IN DUST CLOUDS

Astronomical observations on their own do not give us scientific insight into the universe. These observations only come alive when compared with predictions of a theoretical model. In bidding for observing time on a telescope, we select a question of interest, we summarize what is known about it, and we make theoretical predictions of what the new observations might yield. If the question posed seems interesting enough, the time allocation committee lets us loose on the telescope. We analyze the results and then turn again to the theoretical models to make sense of them. Often this leads to new questions.

The key to understanding infrared emission from stars, interstellar material and galaxies is to understand how radiation flows through a cloud of dust. Optical and ultraviolet light are absorbed by grains of dust, which heat up and then radiate infrared light in all directions. That infrared light may itself be absorbed by other grains. The light can also be scattered or reflected by grains, and this scattered light also contributes to the heating of other grains, so this is a highly complex process. The growth of infrared astronomy in the 1960s and 1970s generated a growing need for theoretical models of infrared sources. These tended to be quite simplistic at first, assuming some simple dependence of dust grain temperature on distance from the source and treating the resulting emission from the grains as a simple superposition of emission from all the different grains, with no allowance for possible absorption of the radiation farther out in the cloud. At the time, it was natural to try to explain the phenomena in the simplest possible way. But models have to be self-consistent, so we really do have to address how radiation flows through dust.

In the late 1970s, Harold Yorke[45] and Chun Ming Leung[46] did pioneering work on computer models for infrared emission from dusty H II regions, which were the first attempts to follow in detail the flow of infrared radiation through dust clouds.[47] My postdoctoral researcher Stella Harris and I used a computer code of this type to model the individual infrared spectra of all the circumstellar dust shells in the AFCRL survey with high-quality data.[48] We used the pioneering stellar observations of Harold Johnson and his group, as well as data on long-period Mira variable stars that had been accumulated by amateur observers. Gillian R. Knapp later used our dust shell models to estimate how much mass each of these stars is ejecting per year and showed that this increases strongly with the stellar luminosity.[49] As stars ascend the red giant branch, becoming ever more luminous, the rate of mass loss increases and the star's atmosphere becomes more and more unstable. Eventually we see the dramatic ejection events that give rise to planetary nebulae like the Ring Nebula, the last remnants of the star's atmosphere being thrown off. The dense, hot core of nuclear ash, typically composed of carbon, nitrogen and oxygen, remains behind at the centre of the nebula and becomes a white dwarf star.

MODELS FOR INTERSTELLAR DUST GRAINS

Interstellar dust plays a crucial role in influencing the physics and chemistry of the interstellar material. Almost all elements heavier than helium in the interstellar gas are tied up in grains, except for some of the carbon and oxygen in the form of carbon monoxide gas. There are dozens of other molecules present in the interstellar medium, mainly compounds of carbon, nitrogen and oxygen, but they make up a negligible proportion of the abundance of those elements.

The dust in the interstellar medium of the Galaxy absorbs about one-third of the starlight in the Galaxy. Except for those close to a star, the dust grains tend to reach a temperature of 10–50 K and reradiate their energy at far-infrared wavelengths. During the 1960s and 1970s, the observational properties of interstellar dust grains began to be better understood.[50] In 1979, Blair Savage and John Mathis summarized the observational properties of interstellar grains from ultraviolet to infrared wavelengths, combining the results from many observational studies of the previous decade.[51] The ultraviolet extinction properties were derived from a series of U.S., Dutch and European astronomical satellites. The key new result in the ultraviolet was a broad peak of extinction centred on 0.22 microns, believed to be caused by small graphite particles. In the visible range, the variation of extinction with wavelength approximately followed the inverse-wavelength law found by Whitford, and this behaviour continued into the near-infrared band. A narrow feature at 3.07 microns found towards dense gas clouds was attributed to water ice. Far more ubiquitous was the 9.7-micron silicate feature and a weaker, broad silicate feature at 18–22 microns seen in the infrared spectra of many stars. Carbon stars do not show these silicate features but instead have a narrow feature at 11.2 microns caused by grains of silicon carbide. In addition, many locations showed the 'unidentified infrared bands' at 3.3, 3.4, 6.2, 7.7, 8.7 and 11.3 microns, so it was clear that interstellar grains comprised silicates, graphite or amorphous carbon, and some other ingredient that accounted for the unidentified infrared bands. The main ingredients of the dust between the stars are essentially finely ground sand and soot.

It was obviously of interest to find a detailed model for interstellar grains with these various ingredients to fit the observed dust extinction law. Such a model would give far greater predictive power, allowing an estimate of what would happen at wavelengths where extinction and other interstellar dust properties had not yet been measured. John Mathis and his colleagues produced a good fit to the average extinction curve with a mixture of graphite and silicate grains, with the grains having radii between 0.005 and 0.25 microns and with far more of the smaller grains than of the larger ones.[52] So this is a much finer dust than we encounter on Earth, where particles are more typically tens or hundreds of microns in size. From 1984 onwards, Bruce Draine and his collaborators developed more sophisticated versions of this interstellar dust model, with refined optical properties for the different grain types.[53] These models now gave the dependence of absorption, scattering and extinction as a function of wavelength, from x-rays to far-infrared wavelengths.

The different grain components may have their origins in different types of stars. The larger amorphous grains of carbon or silicates probably originate in outflows from red giant stars,[54] but at least some of the smaller grains could have been formed in supernova explosions from massive stars. One of the places where we can see these different sources of grain material is in meteorites that have fallen to Earth. Meteorites are generally remnants of cometary material which have been formed in the outer reaches of the Solar System and therefore hold clues to the history of the interstellar material from which the

Solar System was assembled. Analysis of meteorites shows that they contain many small 'inclusions', and the composition of these shows that they originate in different types of stars. Some of them are clearly debris from nearby supernova explosions.

One key ingredient not present in the dust grain models of the 1970s and early 1980s was the component responsible for the unidentified infrared bands, which had still not been identified. Steve Price had commented to me that the 11-micron background radiation in his AFGL rocket flights seemed to him to be surprisingly high. What did I think it was? I suggested, wrongly, that it must be integrated radiation from circumstellar dust shells around stars since this was the main type of source being seen in the AFGL survey. In the early 1980s, Tom Soifer told me about extended mid-infrared radiation being found around bright stars by Kristen Sellgren which could not be from dust grains in thermal equilibrium with the starlight.[55] Normally dust grains absorb visible and ultraviolet light and then adjust their temperature so that they radiate exactly at the rate at which they are absorbing energy. Kristen Sellgren showed that the dust grains responsible for the diffuse infrared radiation were at temperatures far too high to be in balance with the radiation they were absorbing.

The explanation for these anomalies, and for the unidentified infrared bands, was finally given by the French astronomers Alain Leger and Jean-Loup Puget, and by the American laboratory experimentalist Louis Allamandola and his group, in 1984–5. The unidentified infrared bands are caused by large molecules of 50 to 100 atoms with a benzene-like (hence 'aromatic') structure. These molecules are known as *polycyclic aromatic hydrocarbons* (PAHs).[56] On Earth these molecules are found in the soot in car exhaust pipes. There are several lines of circumstantial evidence that point to this identification. The emission cannot be thermal radiation from a dust grain in equilibrium with the light from the star, so a large molecule must be responding to a single ultraviolet photon by becoming highly excited and then cooling down. Secondly, it had been established from careful observations of planetary nebulae that the fraction of the total infrared energy that is emitted through these features is closely correlated with the amount of available carbon, proving that the material responsible for the features is carbon-rich. The molecules have to be extremely stable to survive the intense radiation they experience. The similarity of the

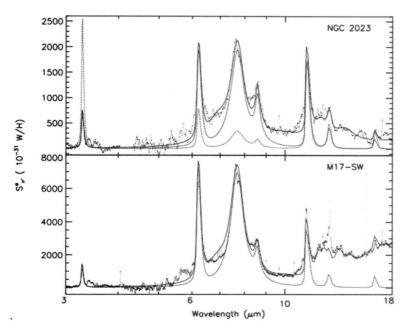

Figure 4.10. Model for the 'unidentified' infrared bands based on a mixture of PAHs studied in the laboratory (Allamandola and Hudgins 2003), compared with the observed spectra from the planetary nebula NGC 2023 and the star-forming cloud M17.

features in different astronomical sources implies that a single class of chemical species is responsible. And crucially, laboratory studies of PAHs in the infrared showed that they are capable of mimicking the wavelengths of the unidentified features. Louis Allamandola, of the NASA Ames Research Center, Mountain View, California, and his colleagues have studied over 60 different types of PAHs in the laboratory and shown how different combinations of them can match the infrared spectra seen in star-forming regions and planetary nebulae (Figure 4.10).[57]

In a series of interesting laboratory experiments that could be relevant to the origin of life, Allamandola and his colleagues have shown that mixtures of PAHs and ices of the kind found in interstellar clouds and in comets (for example, water, carbon monoxide, carbon dioxide, methane, ammonia and methanol), when subjected to ultraviolet radiation, generate even more complex molecules and structures. When these organic residues are dissolved in water, large, hollow droplets up to 10 microns in size are found. These droplets have a composition similar to the types of molecules that make up cell membranes. Here is an intriguing hint that processes in interstellar clouds could have gone a long way towards providing the broth of prebiotic molecules and the type of membrane needed to form living cells. These materials could have been delivered to Earth in the deep freeze of cometary ices during the early phases of planet formation, when the Earth is known to have suffered episodes of major asteroidal and cometary bombardment.

EMERGENCE OF INFRARED ASTRONOMY OUTSIDE THE UNITED STATES

The emergence of infrared astronomy in the 1960s and 1970s was very much dominated by U.S. astronomers, with the pioneering work of Harold Johnson, Frank Low and Gerry Neugebauer leading the way. However, groups were beginning to develop in many other countries, including France, the United Kingdom,[58] Australia[59] and South Africa. Harry Hyland used an infrared photometer at Mount Stromlo in Australia in the early 1970s, and Dave Allen developed an infrared photometer at the Anglo-Australian Telescope in 1978, the first of a series of infrared instruments there. Jim Ring's group at Imperial College London developed instruments for near-infrared astronomy and built a lightweight infrared telescope, the Infrared Flux Collector, on Tenerife in 1972. The Imperial College programme culminated in Bob Joseph's work on the role of interactions and mergers between galaxies in explaining the high infrared luminosity seen in some galaxies[60] and the detection of molecular hydrogen in two merging galaxies.[61]

The period from 1960 to 1980 saw astronomy pushing out from the visible band into the near- and mid-infrared, working mainly from mountaintops and using a set of wavelength windows between 1 and 20 microns. The main scientific discoveries were of dust shells surrounding dying stars, of stars in the process of formation, and of the nature of interstellar dust. As we will see in the next chapter, the 1970s also saw the emergence of submillimetre astronomy, with work at the much longer wavelengths of 100–1000 microns. At the longer end of the wavelength band, some work could be done from mountaintops, but mostly it was necessary to get above as much of the Earth's atmosphere as possible by using aircraft and balloons.

5

Birth of Submillimetre Astronomy: Clouds of Dust and Molecules in Our Galaxy

We have seen that at near- and mid-infrared wavelengths there are a few wavelength windows accessible from mountaintop observatories, at 1.25, 1.65, 2.2, 3.5, 5, 10 and 20 microns. In the previous chapter, I described the advances that came with pushing out into these near- and mid-infrared windows with ground-based telescopes and with telescopes on rockets. Beyond 20 microns, carbon dioxide and especially water vapour in the Earth's atmosphere make astronomy from the ground almost impossible until we reach a few submillimetre windows at 350, 450, 850 and 1250 microns (Figure 5.1). In this chapter, I describe how ground-based submillimetre astronomy in these windows developed and how far-infrared and submillimetre astronomy was opened up using aircraft and balloons. The cosmic landscape these wavelengths opened up was one of dense clouds of dust and molecular gas, and of new stars being formed in these clouds.

SUBMILLIMETRE ASTRONOMY

The father of far-infrared and submillimetre astronomy was undoubtedly Frank Low, who in 1961 had invented the gallium-doped germanium bolometer and the Low dewar.[1] In the previous chapter, I discussed how Low and his collaborators applied this detector to 10- and 20-micron observations of stars and of clouds of cool gas where new stars are forming. In the late 1960s, Frank Low began to work at genuinely far-infrared wavelengths using a 12-inch telescope aboard a Lear executive jet (Figure 5.2).[2] His first attempts at airborne astronomy had begun soon after his move to Arizona in the 1960s,

I have made extensive use of the historical review by Low, Rieke and Gehrz (2007).
There are valuable reviews by Zuckerman and Palmer (1974), Morris and Rickard (1982) and Young and Scoville (1991).

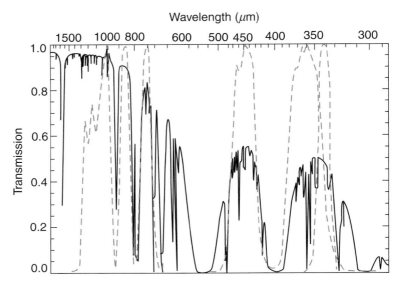

Figure 5.1. Atmospheric transmission at the summit of Mauna Kea as a function of wavelength (full curve, top scale) in the submillimetre band, 300–1500 microns. Also shown, as broken curves, are the response functions of filters used in the submillimetre observing windows.

when with Carl Gillespie he observed the Sun at a wavelength of 1 millimetre (1000 microns) from a twin-jet Navy bomber. The Lear jet was an early executive jet, and the telescope was installed in the emergency door opening. The jet flew at 50,000 feet, and the cabin air pressure was so low that the pilots and astronomer operating the telescope needed full breathing equipment. To subtract out the foreground infrared emission from the residual atmosphere, Low installed a tilting secondary mirror operated by a magnetic drive. This could be used to 'chop' between the target source and a nearby region of blank sky. George Aumann was the first of Low's graduate students to work in the Lear jet, and he, Gillespie and Low measured the temperatures of Jupiter and Saturn, demonstrating that they are warmer than would be expected if they were heated purely by the Sun.[3] These giant planets require their own internal heating source, mainly heat left over from their formation. In 1968, Frank Low and Wallace Tucker estimated the far-infrared background radiation from all the galaxies in the universe, predicting that the background from these galaxies would peak at wavelengths beyond 50 microns and that the intensity would be comparable to that from 'other' (i.e., optical) galaxies, a prediction that turned out to be surprisingly close to the truth, though

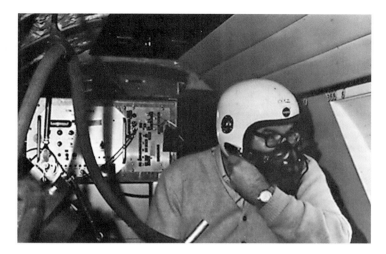

Figure 5.2. Frank Low in the Lear jet airborne observatory. The back end of his infrared telescope can be seen at the lower left.

it would take 30 years to demonstrate this.[4] A very influential study by Low and Aumann in 1970 showed far-infrared spectra of several well-known galaxies,[5] and this pointed the way to the great discoveries of the *Infrared Astronomical Satellite* in the 1980s.[6]

Even before Frank Low began his far-infrared and millimetre observations from the Lear jet, a far less well-known submillimetre pioneer, John Bastin, was carrying out millimetre observations of the Sun and Moon from the roof of the physics department building at Queen Mary College (QMC) in the East End of London.[7] In 1969, Bastin and his group announced the first detection of 1.2-millimetre emission from the Crab Nebula, the remnant of a supernova that exploded in 1054 AD, an event that had been noticed by Chinese astronomers of the time.[8] Peter Clegg and his colleagues detected millimetre radiation from the Orion Nebula as well.[9] In 1969, Bastin and his group made the first genuinely submillimetre measurements of the Sun, at 860 microns.[10]

Peter Ade commented to me:

> John Bastin was the first person to compute (by hand), from the known lab measurements of the water and oxygen absorption lines, the expected transmission of the atmosphere at these wavelengths.... He reported his findings at a conference in the early 1960's where he stated that from a good high altitude site one could make observations at submm wavelengths if we viewed through the window regions between the strong absorptions. At this point Frank Low stood up

and said that this was not possible and that he was mistaken in his calculations! However, having produced the (submillimetre) filters I was able to demonstrate that he was right – as we easily detected and mapped the Sun and Moon from the Pic-du-Midi Observatory and made tentative measurements of the planets and Orion. Therefore technically I believe that the QMC group were the first to knowingly make submillimetre observations.

The key people in stimulating the Queen Mary College group to branch out into wider areas of astrophysics were John Beckman, who was to lead the QMC microwave background experiment (Chapter 6), and Peter Clegg, who was to play a key role in the day-to-day leadership of the *IRAS* mission (Chapter 7) and led the work with the Long Wavelength Spectrometer instrument on the *Infrared Space Observatory* (Chapter 10). My first involvement with the group came in 1972, when Peter Clegg invited me to join Peter Ade and himself on an observing run at the Kitt Peak millimetre wave observatory. It was a wonderful experience to drive through the Arizona desert to Kitt Peak mountain and spend a week observing there (Figure 5.3). The 11-metre telescope dedicated to millimetre-wave astronomy had been designed by Frank Low during his spell at the National Radio Astronomy Observatory (NRAO) and was located on a ridge separated from the main optical telescopes. We worked in shifts, night and day, with wonderful views over the Arizona desert by day and the lights of distant Native American villages by night. The night sky was spectacular: the colours of the stars, the blaze of the Milky Way. We slept in trailers that smelled of skunk, and we had to be careful of snakes and scorpions.

The detectors and dewar being used by the Queen Mary College group were continually developing. By the time of these Kitt Peak runs, we had graduated from an indium antimonide detector to a composite germanium bolometer that had been developed by the French millimetre-wave physicist Noel Coron. The dewar was originally a glass cryostat, like a double vacuum flask, but in common with most workers in the field, we switched to a Frank Low metal dewar. Peter Ade carried the dewar and detector system in his hand luggage across the Atlantic, something which would probably not be allowed today. The Queen Mary College speciality was the filters designed by Peter Ade that defined the waveband and excluded, as much as possible, water-vapour emission lines. I was very keen that we should observe extra-galactic objects, but although we detected several new extragalactic and Galactic sources at 1 millimetre, our sensitivity was not quite good enough for a major breakthrough.[11] A true breakthrough would require

Figure 5.3. The Queen Mary College group at the Kitt Peak Millimetre Wave Telescope in 1976. From the left: Peter Clegg, Ian Robson, the author and Peter Ade.

completely new, dedicated telescopes and significant instrumental advances. In the period 1968–72, while the Queen Mary College group was struggling to observe at wavelengths around 1 millimetre, the U.S. infrared astronomers were going through the extraordinary explosion of 2–20-micron astronomy summarized in the last chapter.

The 1960s and 1970s also saw the emergence of several U.S. groups working in far-infrared and submillimetre astronomy, at Cornell, Goddard Space Flight Center, Harvard, Caltech, Chicago and Stony Brook, and of a group at University College London.

Martin Harwit at Cornell University had done his PhD thesis on thermal fluctuations in infrared radiation in 1960. He became interested in the possibility of working in infrared astronomy and wanted to observe with cold telescopes from space. He contacted the U.S. Naval Research Laboratory, where a programme of successful rocket observations in the x-ray and ultraviolet bands had been carried out by Herbert Friedman. Harwit spent a year at the NRL and immediately found there was an opportunity to put a cluster of infrared telescopes on a secret Air Force Atlas rocket launch. After they had frantically completed their instrument package, the Air Force delayed their part of the mission, and the entire launch, for a year. Harwit describes what followed:

> When the Atlas flight did take place, it was a total fiasco. On the night of the flight, the launch kept getting further and further delayed. When the maximum four-hour hold time (of the liquid nitrogen dewar) had passed, I requested that the payload be removed from the missile, but the Air Force now needed it as ballast and refused. Finally, the launch took place after a six-hour delay on the launch pad. The telemetry signals showed the last of the liquid nitrogen evaporating just as the rocket lifted off. It was heartbreaking.[12]

The Cornell group switched to smaller Aerobee rockets, which allowed them to control the launch and allowed about five minutes of observation above the Earth's atmosphere per flight (Figure 5.4). Problems with stray light from the Earth, and from the aura of gases and particles that surrounded the rocket even at that altitude, undermined efforts to observe the infrared cosmic background radiation. Jim Houck joined the group after finishing his PhD in solid-state physics at Cornell, and not long after, the group had a series of successful flights with helium-cooled telescopes. They made the first successful measurements of infrared emission from zodiacal dust in the Solar System in 1971 and went on to map the Galactic centre.[13] An important result of these rocket observations was the demonstration that many sources of far-infrared emission were extended beyond the telescope's field of view, so that ground-based, chopped observations with small apertures were missing much of the radiation. The Cornell group also published influential theoretical work on expected infrared lines from molecular hydrogen[14] and on dust cocoons around massive young stars.[15]

Bright, extended emission from the centre of the Galaxy was found at a wavelength of 100 microns by Bill Hoffmann and his group in the 1960s. Hoffmann, working at the Goddard Institute for Space Studies in New York, had become interested in predictions that interstellar dust would be heated by starlight. He built a balloon gondola carrying a 1-inch telescope placed within a liquid-helium dewar, with a Low germanium bolometer detector (Figure 5.5). The first flights were disappointing, detecting the Moon but no other objects. Then, in 1969 they detected the Galactic centre and announced: 'We believe this to be the first object outside the solar system that has been detected in the region of the spectrum from 25 microns to 1000 microns.'[16]

Figure 5.4. Martin Harwit working on the payload for an Aerobee rocket launch.

Soon after this, Hoffmann's group made a pioneering 100-micron map of part of the Galactic plane with the Goddard Institute for Space Studies balloon-borne telescope.[17] At the balloon's altitude of 27 kilometres, over 99% of the 100-micron radiation from the universe reaches the telescope, compared with essentially zero at the ground. Hoffmann's group detected 72 sources, of which 60 were identified with radio sources (regions of ionized hydrogen), bright nebulae, dark nebulae or infrared stars.

The University College London group, led by Dick Jennings, announced the first results from their balloon programme at 40–350 microns in 1972. Their instrument was designed to isolate the far-infrared part of the spectrum, and they launched their telescope from the balloon facility at Palestine, Texas. Over the next few years, they mapped many of the brightest sources of far-infrared radiation in the Galaxy, which are associated with giant clouds of dust and molecules where clusters of massive stars are forming.[18]

Other groups were also working at a variety of far-infrared and submillimetre wavelengths. The group based at the State University of New York at Stony Brook announced detections of Venus, the star-forming cloud M17, the Galactic centre and the nearby galaxy

Figure 5.5. Bill Hoffmann about to launch his balloon gondola.

M82 from ground-based observations in the 350-micron window.[19] Giovanni Fazio's group at Harvard announced results from their balloon programme on Orion at 69 microns.[20] And in 1976 NASA's *Kuiper Airborne Observatory* started work with a 0.9-metre telescope aboard a C-141 aircraft. The Chicago group quickly used this to measure an impressive spectral-energy distribution from 28 to 320 microns of the Seyfert galaxy NGC 1068.[21] The Caltech group was also beginning to work from Mount Palomar at 1 millimetre and announced detections of several new Galactic and extragalactic sources.[22]

In 1980, George Rieke, Frank Low and their collaborators completed a classic study of M82 and NGC 253, the two nearest bright infrared galaxies.[23] They found that M82, with an infrared luminosity of 40 billion solar luminosities, must be forming stars at a rate at least ten times higher than our own Galaxy. The mass of gas in its central regions, 250 million solar masses, could sustain this rate of star formation for no more than 60 million years, rather a short episode in the lifetime of a galaxy, and so the term *starburst* was coined. These bursts of star formation could account for the large infrared and radio luminosity of M82, the large populations of red giants and supergiants, and the large amount of ultraviolet and x-ray radiation. A starburst produces fireworks at all wavelengths, although the bulk of the energy output is in the far infrared.

MOLECULAR-LINE ASTRONOMY

So far I have talked about the growth of submillimetre *continuum* sources, in which the radiation is spread smoothly over all submillimetre wavelengths. During the 1970s another very important development emerged, millimetre molecular-line astronomy, where we see radiation only at the characteristic wavelengths of interstellar molecules. This was made possible by the combination of the NRAO 11-metre millimetre-wave telescope on Kitt Peak, Arizona, and the semiconductor-based 'Schottky diode' receiver, which could provide sufficient spectral resolution to detect millimetre emission lines. The key first discovery was by Bob Wilson, Keith Jefferts and Arno Penzias, who detected the carbon monoxide (CO) molecule at a wavelength of 2.6 millimetres in the Orion Nebula in 1970.[24] The spectral lines from atoms arise from jumps between the different energy levels that the electrons can occupy around an atomic nucleus. In the case of a molecule with, say, two atoms in a dumbbell configuration, there are also different possible energy levels for the rotation of the molecule, and jumps between these levels produce microwave or submillimetre radiation. In CO, the transition from the first excited level to the lowest-energy ground state (called the 1–0 transition) results in radiation of wavelength 2.6 millimetres.

The discovery of millimetre emission from carbon monoxide triggered an explosion of discoveries by Penzias, Wilson and their collaborators, all using the 11-metre NRAO millimeter telescope at Kitt Peak, Arizona.[25] Within a year, nine new simple molecules had been discovered.[26] Most interstellar molecules form on the surfaces of interstellar dust grains, within dense clouds of gas and dust. The main requirement for molecule formation is that the dust provide sufficient obscuration to shield the inner parts of the cloud from ultraviolet light, which tends to destroy molecules very quickly, converting them back to their constituent atoms.

Although these were the first interstellar molecules detected at millimetre wavelengths, the very first interstellar molecules to be identified, in 1940, had been methylidene, CH, and the cyanogen radical,

CN, which have strong transitions at optical wavelengths.[27] The rise of radio astronomy led Iosef Shklovsky in 1952[28] and Charles Townes in 1955[29] to list molecules that might have detectable radio transitions. The first of these to be detected was the hydroxyl radical (OH), which was detected in 1968 in absorption looking towards the radio source Cas A, the remnant of a supernova that exploded about 11,000 years ago in the constellation of Cassiopeia.[30] Townes and his group decided to search for interstellar ammonia and succeeded.[31] Interstellar formaldehyde was detected at a wavelength of 2 centimetres in 1969.[32] These three molecules had the great advantage that they could, with the right receivers, be studied with existing radio telescopes.

Much of the early work on molecular-line astronomy was about detecting new molecules and new transitions, but another key benefit of working with radiation from spectral lines, the ability to measure velocities through the red- or blue-shifting of the line, began to be exploited. In 1975, Nick Scoville, Phil Solomon and Arno Penzias reported a detailed study of the molecular cloud Sgr B2 near the Galactic centre.[33] They mapped the cloud in molecular lines of carbon monoxide, carbon monosulfide and formaldehyde, and found that the cloud contained a dense core surrounded by a more diffuse envelope of gas 150 light years across. They estimated the mass of the cloud in three independent ways: from the CO column density (the number of CO molecules in the line of sight through the cloud), from the dynamics of the cloud, and from far-infrared measurements of its dust mass. The three methods agreed reasonably well, and the researchers estimated that the total mass of the molecular cloud was about three million solar masses. They found that the temperature of the molecular gas was about 20 K, similar to that of the dust grains, so that gas and dust are sharing heat energy equally. They found no trend of velocity with position in the cloud, which suggested that the cloud was in large-scale, turbulent motion rather than in some ordered rotation.

Molecules are also found in the dust shells around dying stars. The archetypal carbon star with a thick circumstellar dust shell is IRC +10216, found in Neugebauer and Leighton's Two Micron Survey. The shell is rich in molecular gas, mainly hydrogen. The next most abundant molecule in this shell of gas and dust is carbon monoxide, and over a dozen other molecules have also been detected in it.[34] These molecules are believed to form in complex chemical interactions in the atmosphere of the star. The chemistry then freezes out as the spherical shell of molecular gas expands away from the star and the pressure drops. Far fewer molecules are seen in the shells around oxygen-rich stars.

In its early years, molecular-line astronomy was confined to the wavelength range 2–3 millimetres accessible to the Schottky diode receiver on the 11-metre Kitt Peak telescope. The key instrumental development that allowed molecular-line astronomy to progress to genuinely submillimetre wavelengths was the invention of the liquid-helium-cooled indium antimonide 'hot electron' bolometer by Tom Phillips at Bell Labs in 1973.[35] In the same year, Phillips, Jefferts and Peter Wannier announced the detection of the CO 2–1 line at 1.3 millimetres in Orion.[36] The line was so strong that it was first detected by observing through the canvas side of the telescope dome during a rainstorm. The CO 2–1 line had the advantage that you could see farther into dense molecular clouds than for the more easily observable CO 1–0 line at 2.6 millimetres. Soon after this, Phillips moved from Bell Labs to Queen Mary College London, and had a great impact on the group there during his two-year stay. A few years later, in 1979, he invented a new submillimetre receiver based on the superconducting properties of a semiconductor at temperatures near absolute zero, the 'SIS' receiver (short for semiconductor-insulator-semiconductor). This revolutionized submillimetre spectroscopy and has become the standard receiver in the field, in use today on the *Herschel Space Observatory* and at the ALMA Submillimetre Array (see Chapter 12).

In 1976, Tom Phillips and I reviewed progress in molecular-line astronomy for the science magazine *New Scientist*.[37] We were able to report that 32 interstellar molecules had been discovered already. Today the figure is over 140! Almost all are organic molecules (i.e., carbon-based), and they demonstrate the richness of the chemistry in interstellar clouds. The largest individual molecules detected to date have 13 atoms. Some molecules can only be detected at submillimetre wavelengths, but the main interest in pushing towards these wavelengths is to detect the more excited transitions of simple molecules such as carbon monoxide and hydrogen cyanide that can be used as diagnostics of physical conditions in interstellar clouds.

Sometime in late 1974, I circulated a memo to the Queen Mary group entitled 'Proposal for a large millimetre telescope', floating the idea of a dedicated U.K. millimetre telescope. The two options I discussed were a 30-metre telescope to operate down to 1 millimetre and a 10-metre telescope operating down to 300 microns, the latter requiring a site on Mauna Kea in Hawaii. In 1975, the United Kingdom carried out a review of radio astronomy. Tom Phillips received a phone call from Martin Ryle saying that the Manchester radio astronomers were proposing to construct a submillimetre telescope by upgrading the famous Jodrell Bank radio telescope. Ryle suggested that Phillips should put forward a more realistic proposal at a better site. Phillips took up this idea enthusiastically and submitted a detailed proposal for review. He has described to me what followed:

> At the committee meeting both Jodrell Bank and I presented our cases and then I asked the Jodrell Bank team in the question time what their intended science was, since it hadn't been given. They said they wanted to detect new molecules in the Galactic Center. Knowing that the Galactic Center did not rise very high over the horizon at Kitt Peak, which is a lot further south than Manchester, I asked how high the Galactic Center is in transit as viewed from Manchester. There was some discussion amongst the Manchester people before they replied that maybe they would have to cut down the trees. This settled the matter as to which proposal should be followed up on.

Later in 1975, the Millimetre Steering Committee, chaired by Tom Phillips, concluded that it would be possible to construct a telescope of 15 metres diameter capable of operating down to 750 microns.[38] It would take 12 years to bring this idea to fruition. Finally, in 1987, the James Clerk Maxwell Telescope on Mauna Kea, Hawaii, saw first light (see Chapter 9). It would become the dominant instrument worldwide in submillimetre astronomy.

In the late 1970s, Pat Thaddeus and his group started to survey the entire Milky Way with the carbon monoxide 1–0 line using a small (1.2 metres) telescope on the roof of Columbia University in New York. This would be a heroic enterprise, taking over 20 years. The idea that millimetre-wave observations could be carried out at sea level in the middle of a huge city seemed startling at first, though John Bastin had done observations in the East End of London during the 1960s, albeit mainly of the Sun and Moon. During the cold winter months, the level of water vapour in the atmosphere above New York was low enough to permit millimetre observations. Between 1979 and 1986, Thaddeus and his group surveyed nearly one-fifth of the sky, using a second telescope at Cerro Telolo in Chile to cover the Southern Hemisphere. I remember seeing Columbia University's rooftop observatory during a visit there in the early 1980s, with its excellent view across to Harlem. The final results of this first survey were reported in 1987.[39] The Columbia group then embarked on a higher-resolution, more sensitive survey, which was completed in

2001.[40] These surveys gave not only the location of the molecular clouds in the Galaxy, tracing out the Galaxy's spiral arms, but also the detailed motions of this gas.

In his PhD research with Thaddeus, Leo Blitz mapped the molecular gas around the known OB associations in the Milky Way using the 2.6-millimetre line of CO.[41] These groups of massive, hot stars of spectral type O and B signpost the regions where massive stars are forming in the Galaxy and were already known to be the locations of giant regions of ionized hydrogen from radio observations. Blitz's survey demonstrated that massive star formation takes place in giant molecular clouds. The OB stars and ionized regions corresponded to locations where newly formed massive stars were breaking out of the molecular cloud. Most of these regions were also known from the balloon surveys of the Goddard, Harvard and University College groups to be strong sources of far-infrared radiation. The masses of these molecular clouds ranged from one thousand to several hundred thousand solar masses, but they had rather similar far-infrared spectra, peaking at around 100 microns.

MOLECULES IN EXTERNAL GALAXIES

The first observations of carbon monoxide in external galaxies were made in 1975 by Tom Phillips and his group, in the Magellanic Clouds,[42] and by Lee Rickard and his co-workers, who detected the starburst galaxies M82 and NGC 253.[43] Progress in detecting external galaxies was slow at first, and when Mark Morris and Rickard reviewed molecular clouds in external galaxies in 1982 they reported only 18 published detections.[44] The problem was that astronomers did not know which galaxies were likely to be strong sources of molecular emission. The situation changed in 1983 with the launch of the *Infrared Astronomical Satellite*, which surveyed the whole sky at far-infrared wavelengths and detected tens of thousands of bright infrared galaxies (see Chapter 7). When Judith Young and Nick Scoville reviewed progress a decade later, over four hundred galaxies had been mapped and detected in CO.[45] They concluded that the luminosity in the carbon monoxide line is a good measure of the total mass of molecular gas in a galaxy. They found that the latter is proportional to the infrared luminosity, and this in turn is a good measure of the rate of star formation in a galaxy.

The submillimetre band had revealed the existence of dense molecular clouds, with a complex, mainly carbon-based, chemistry.

These were the nurseries where massive new stars were being assembled. Detection techniques in the 1970s and early 1980s were still quite primitive, and only a few external galaxies had been detected through their dust or molecular emission. But another highly significant discovery had been made at microwave wavelengths which was to transform our view of the origin of the universe itself: the cosmic microwave background radiation.

6

The Cosmic Microwave Background, Echo of the Big Bang

The story so far has been of stars shrouded in dust, clouds of gas and dust where new stars are forming, and nearby star-forming galaxies. Now we shift the stage to the whole universe with the discovery of background radiation at microwave wavelengths. The cosmic microwave background (CMB) radiation is the dying whisper of the initial fireball phase of the hot Big Bang universe, and its discovery transformed our understanding of the universe. This background is the dominant form of astronomical radiation at submillimetre wavelengths outside the Milky Way. When astronomers were trying to detect submillimetre sources in the 1970s, nodding the telescope between the source position and a nearby position on the sky to subtract out the emission from the Earth's atmosphere, each telescope beam would detect an amount of CMB radiation far brighter than the source they were trying to detect. However, because they were only interested in the difference between the two measurements, the CMB radiation exactly cancelled and did not affect the observations. In this chapter, I describe the discovery of the radiation and its impact on cosmology.

The story of the discovery of the cosmic microwave background by Arno Penzias and Bob Wilson in 1965 has been told many times.[1] Initially this was a story of short-wavelength radio astronomy and microwaves (centimetre wavelength radiation), but gradually it became clear that the wavelength of peak energy is around 1 millimetre, so while about half the CMB is microwave, half is submillimetre. To understand the significance of this radiation, we have to understand the origin and evolution of the universe itself.

THE EXPANSION OF THE UNIVERSE

In 1924, Edwin Hubble (1889–1953) demonstrated that the Andromeda Nebula, M31, lay far outside our Milky Way Galaxy, thus opening up the universe of galaxies and solving the century-old riddle of whether the spiral nebulae were part of the Galaxy or not. A few years later, in 1929, he demonstrated that the universe is expanding and that the farther a galaxy is from us, the faster it is receding (which came to be known as Hubble's Law). He estimated the recession velocity of the galaxy from the redshifting of its spectral lines, the famous Doppler effect, discovered by the Austrian physicist Christian Doppler in 1842. Doppler realized that when a source of waves, for example sound waves, is moving towards us, the pitch or frequency of the waves is increased. When the source is moving away from us, the frequency is reduced. In the case of light, this means that the colour of an optical source is shifted towards the blue end of the spectrum when the source is moving towards us and towards the red end when it is moving away from us. The *redshift* of a galaxy is defined as the fractional increase in the observed wavelength relative to the emitted wavelength, and to first approximation (valid for nearby galaxies) it gives the ratio of the recession velocity to the speed of light. Hubble found that apart from a few very nearby galaxies, all galaxies are moving away from us. More importantly, the redshift, or inferred speed of recession, increases with the distance of the galaxy from us. The universe is expanding away from us, and we appear to be at the centre of this expansion.

Now this apparent central position we occupy in the expanding universe is an illusion. Someone sitting in another distant galaxy would see exactly the same picture. They would see us moving away from them. This universal expansion is a consequence of the remarkable simplicity of our universe. This simplicity is summed up in a principle first proposed by Albert Einstein in 1917, the Cosmological Principle. This proposes that the universe is *homogeneous* (every observer sees the same picture of the universe) and *isotropic* (the picture looks the same in every direction). Einstein had produced his theory of gravitation, the General Theory of Relativity, a year earlier and was now trying to apply it to the universe as a whole. To make things mathematically simple, he assumed the universe satisfied the Cosmological Principle, although at the time there was not a shred of evidence to support this, and quite a few pieces of evidence pointing to a very nonisotropic universe. Amazingly Einstein's guess turned out to be true for the universe to a remarkable precision.

Although Einstein turned out to be right about the isotropy of the universe, his 1917 model for the universe was wrong in almost every other respect. He was trying to make a static model of the universe, and Hubble's discovery soon shattered that. As it turned out, a young Russian mathematician and meteorologist, Alexander Friedmann, had already shown that an expanding universe was to be expected in a simple, smooth, uniform, isotropic (looking the same in every direction) universe obeying Einstein's General Theory of Relativity.

Hubble's 1929 discovery was revolutionary, but at first the Hubble Law, that redshift is proportional to distance, could only be tested for very small redshifts, because these could be measured only for rather bright, nearby galaxies. Even in the 1960s, it was still pretty exciting to detect a galaxy with redshift 0.1 (i.e., spectral lines shifted by 10% towards the red). In theory, a galaxy at the limit of the observable universe, the cosmological *horizon*, would have an infinite redshift.[2] Today we have detected galaxies with redshifts around 8 (i.e., the wavelengths of their spectral lines have been increased by 800%, or multiplied by 9). We think the first galaxies probably formed at redshifts between 10 and 30. As we look out to ever greater distances, and to ever greater redshifts, we are also looking back in time. To set the scale, redshifts of 0.1, 1 and 10 correspond to times 12.4, 6.0 and 0.65 billion years after the Big Bang, respectively, or to 'look-back' times of 1.3, 7.7 and 13 billion years in the past. The advantage of using the redshift to measure distance in the universe is that it's something we directly measure. Distances or look-back times have to be calculated from a model.

If we imagine the expansion of the universe running backwards, then everything would crash together into an instant of infinite density at a finite time in the past. We call this moment of infinite density the initial singularity, or more colourfully in a phrase invented as a term of abuse by Fred Hoyle, the Big Bang. Today we measure the time since the Big Bang as 13.7 billion years, with an uncertainty of only a few per cent. We'll see how we came to this remarkably precise figure in Chapter 8.

BACKGROUND TO THE PENZIAS AND WILSON DISCOVERY

It is strange to recall how speculative the subject of cosmology was in 1965, before the Penzias and Wilson discovery. This was in fact the year I started work as a postgraduate student in cosmology, so I remember it quite vividly. For some scientists, their main interest in cosmology was as a mathematical application of Einstein's General Relativity theory, and there was work on anisotropic models and on the geometrical structure of the initial singularity. There was interest in developing the formalism of observational tests of cosmology, but little in the way of observational data on which to test these ideas. General Relativity itself was not that well established, and there was exploration of alternative theories, particularly in their applications to cosmology. The steady-state theory of Hermann Bondi, Thomas Gold and Hoyle was much in vogue in the United Kingdom, although it had fewer supporters elsewhere.

Rather few people were actually thinking about the early phases of a hot Big Bang universe, which would be dominated in its early stages by radiation rather than by matter. George Gamov and his collaborators had studied this era in the 1940s and 1950s, motivated by Gamov's conviction that all the elements were made in the Big Bang. This idea had been severely undermined by the demonstration that all the elements apart from helium, lithium, beryllium and boron, the four lightest, were manufactured in nuclear reactions in the centres of stars and in supernova explosions. The hot, radiation-dominated phase of the Big Bang remained the best bet to explain the formation of most of the helium in the universe. The Soviet cosmologists led by Yakov Zeldovich had explored the physics of the Big Bang and understood very clearly that an implication of a hot Big Bang phase would be microwave background radiation. They even understood that the best instrument to detect this would be the Bell Labs microwave antenna at Holmdel, New Jersey, which had been built for telecommunications experiments. However, they became discouraged as a result of a misunderstanding of published measurements made with this antenna, which they thought meant that no significant background could exist. Instead Zeldovich's group turned to discussion of cold Big Bang models.

BELL LABS AND PRINCETON

The group that was most actively thinking about the hot Big Bang models in 1965 was based at Princeton. Bob Dicke and David Wilkinson were preparing a microwave antenna to search for microwave background radiation in order to test an alternative cosmological model of Dicke's. And Jim Peebles was calculating the amount of helium produced during the first minutes of a hot Big Bang and was about to publish an estimate of the background temperature of at least 10 K. The

Figure 6.1. Bob Wilson and Arno Penzias in front of the Bell Labs antenna with which they discovered the cosmic microwave background in 1965.

Princeton plans were thrown into disarray when they heard about the results of two young radio astronomers working at Bell Labs.

Arno Penzias and Bob Wilson had joined Bell Labs on completion of their PhDs. They planned to use the microwave antenna at Holmdel, New Jersey, for a series of radio astronomy measurements (Figure 6.1). They wanted to measure the spectrum of the Milky Way at higher frequencies than had been attempted before and wanted to make accurately calibrated measurements of the bright radio supernova remnant Cassiopeia A. However, they immediately ran into a difficulty that was well known to the Bell Labs engineers who had tried to use the antenna. The telescope was surprisingly noisy at microwave frequencies. Penzias and Wilson did everything they could think of to eliminate this noise, taking the telescope apart to make sure the joints were properly connected, cleaning the surface of pigeon droppings (and getting rid of the pigeons), measuring the shape of the antenna response pattern by flying around it in a helicopter while pointing a small transmitter at it. The noise persisted and was steady whichever direction on the sky they pointed at. Gradually Penzias and Wilson became convinced that the 'noise' was background radiation from the universe. The brightness of the radiation corresponded to a blackbody temperature of about 3 K.

It seems that Penzias and Wilson were not aware of the 1948 discussion by Gamov's collaborators Ralph Alpher and Robert Hermann of a residual temperature from the hot Big Bang of 5 K. It was only when Bernie Burke visited Penzias and Wilson and told them of the Princeton programme that they realized the cosmological significance of their microwave background radiation. Their discovery was reported in a very brief article in 1965.[3] An accompanying article by Peebles, Dicke, Peter Roll and David Wilkinson spelled out the cosmological significance of this radiation.[4] Peebles's calculation predicting a 10 K background was never published, though it had already circulated informally as a preprint. Roll and Wilkinson went on to confirm the existence of the background a few months later with the Princeton radiometer.[5] And the Penzias and Wilson paper on Cassiopeia A[6] has the footnote 'The equivalent temperature of the antenna when pointed at the zenith is about 7 K, about 2.3 K of which is due to absorption by oxygen in the atmosphere', which might have been their sole reference to the inexplicable noise had they not been visited by Bernie Burke.

THE DIPOLE ANISOTROPY

Measurements at other radio and microwave wavelengths all obtained the same background temperature of around 3 K, so the cosmological interpretation, which required an exactly blackbody spectrum, was supported. And the isotropy of the radiation, the fact that it looked the same whichever direction on the sky you looked, was confirmed by many experiments. Eventually ground-based experiments by Bruce Partridge and others, balloon-based experiments by Paul Henry in the United States and Francisco Melchiorri in Italy, and, crucially, maps of the sky made from a U2 aircraft by George Smoot and his colleagues,[7] showed that there is a tiny anisotropy in the cosmic microwave background (CMB) radiation. In one direction on the sky it is one part in a thousand hotter than average, in the opposite direction it is the same amount cooler, and it varies smoothly across the sky between these two extremes (Figure 6.2). This is known as the *dipole anisotropy* and is caused by the motion of our whole Milky Way Galaxy through the cosmic frame. Later, with galaxy samples based on the *IRAS* survey (Chapter 7), we were able to demonstrate the cause of this motion.

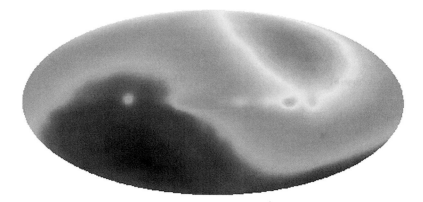

Figure 6.2. The cosmic microwave background dipole anisotropy, mapped by the *COBE* satellite (see Chapter 8). The background appears slightly hotter to the upper right, in the direction towards which the Galaxy is travelling, and slightly cooler to the lower left. Some contamination from the Milky Way can be seen across the middle of the plot.

It is somewhat paradoxical, in the framework of the General Theory of Relativity, that there should be an absolute frame of reference, the one in which the universe is seen to be expanding perfectly isotropically. In the nineteenth century, scientists believed that there must be some kind of medium through which light waves propagated, the *aether*, and there were subtle experiments designed to try to detect the Earth's motion through the aether. The failure to detect the expected aether drift was finally explained within Einstein's Special Theory of Relativity, in which the speed of light in a vacuum is the same for all observers experiencing no forces and in uniform motion with respect to each other. So, in Special Relativity, there is no absolute frame and there is no aether drift. With the CMB dipole anisotropy, we have a new kind of aether drift. Because of the special kind of universe we find ourselves in, there is again an absolute frame of reference, the one from which the expansion looks isotropic.

HISTORY OF THE HOT BIG BANG

The microwave background radiation is the relic of the hot, radiation-dominated early phase of the universe's evolution. While the density of matter declines as the cube of the size of the universe as it expands, because of the change in volume, the density of radiation declines as the fourth power of the size, because there is an extra factor resulting from the redshifting of the photons. This means that however small the amount of radiation in the universe is today, as we extrapolate back to the past and the size of the universe tends to zero, the energy density of the radiation will eventually dominate that of matter. Radiation is the dominant form of energy for the first 30,000 years, but for another 350,000 years, as the temperature drops from 10,000 to 3000 K, radiation still controls the motion of matter through the powerful scattering effect of free electrons. Only when the temperature is cool enough for electrons to combine with protons to make neutral atomic hydrogen, at the *epoch of recombination*, does radiation start to stream freely through the universe, eventually hitting our telescopes as the microwave background radiation, with the redshift changing the temperature from 3000 K to 2.7 K. At the epoch of recombination, matter is freed of the controlling effect of the radiation and can collect together under the influence of gravity and start to make stars and galaxies.

All this history had been well understood for many years before Penzias and Wilson's discovery, and it was also realized that when the temperature was around 10 billion degrees protons and neutrons would have fused together to make helium. It was now worthwhile to calculate the cosmological nucleosynthesis in much greater detail, and as a postdoctoral project, working with Willy Fowler and Fred Hoyle, Bob Wagoner recalculated this nucleosynthesis using a network of several hundred nuclear reactions.[8] He confirmed that as well as helium, small amounts of deuterium (heavy hydrogen, with an extra neutron in its nucleus) and lithium would be produced, but little else. The good agreement of the abundances of helium, deuterium and lithium with the predictions of Big Bang nucleosynthesis calculations was one of the triumphs of the Big Bang theory.

DISTORTIONS TO THE BLACKBODY SPECTRUM?

Although the broadly blackbody form of the cosmic microwave background spectrum had been demonstrated in microwave measurements at a wide range of wavelengths, it was important to test this at shorter wavelengths. In 1969, while discussing millimetre and submillimetre astronomy and mainly focusing on 1.2-millimetre observations of the Moon, Peter Clegg and his colleagues discussed the possibility of measuring the cosmic thermal 3 K background at wavelengths of 1.9 and 1.3 mm,[9] close to the peak of the background spectrum. They eventually achieved this in a balloon-borne experiment in 1974.[10] They showed that the spectrum does indeed show the turnover at wavelengths below 1 millimetre expected for a 2.7 K blackbody, which was the consensus temperature from ground-based measurements. Similar results were found a year later by a group at Berkeley led by Paul Richards, but neither group could unequivocally confirm a blackbody spectrum for the background. In a later experiment, Paul Richards's group claimed in 1979 the surprising result that 'the data, however, do not fit a simple Planck curve with a single temperature',[11] and this distortion from a blackbody spectrum appeared to be confirmed by a later rocket-borne collaboration between the Berkeley group and a group at Nagoya in Japan.[12] Such a distortion was unexpected according to Big Bang cosmology, and there were attempts to understand it in terms of extra emission from dust illuminated by an early, pregalactic, population of stars. The spectral distortion was eventually shown not to exist by the *COBE* satellite team in 1990 (see Chapter 8).

Herb Gush, at the University of British Columbia, had been trying to make rocket measurements of the spectrum of the cosmic microwave background since the early 1970s. After many failures, he successfully measured a blackbody spectrum in 1990, a couple of months after the *COBE* announcement.[13]

SMALL-SCALE ANISOTROPIES IN THE COSMIC MICROWAVE
BACKGROUND: THE NEED FOR DARK MATTER

The early measurements of the cosmic microwave background showed that it is approximately isotropic, the same in every direction. By 1977, the dipole anisotropy resulting from the Galaxy's motion through the cosmic frame had been found. But the background radiation ought also to show small-scale structure caused by fluctuations in density corresponding to the first steps towards the formation of galaxies and clusters of galaxies. If the radiation were perfectly smooth, it would mean that the universe would never develop any structure. There would be no galaxies, no stars, no us. In the late 1970s, ground-based experiments attempted to measure these anisotropies, without success. As ever stronger limits were set, theories for the origin of structure in the universe, galaxies and clusters of galaxies, had to be repeatedly modified. From about 1980 onwards it became clear that a universe containing just ordinary baryonic matter (protons and neutrons) and radiation could not be consistent with the low level of anisotropies implied by the observations. The most distant galaxies known at this time had to have formed within a few billion years of the Big Bang. For gravity to amplify density fluctuations to the point where these galaxies could have formed, the density fluctuations would have already had to have an initial amplitude at least 0.1% of the average density of matter at the epoch of recombination. Observational limits on the fluctuations in baryonic matter were already much smaller than this, so there had to be some other form of matter present in the universe that had escaped the control of radiation at a much earlier time. Much larger fluctuations could have developed in this matter component without imprinting any pattern on the microwave background radiation.

From this emerged the widespread belief among cosmologists that the matter in the universe must be mainly in the form of some new kind of *dark matter*. The existence of dark matter in the outskirts of galaxies had been demonstrated in the 1970s by Vera Rubin and her collaborators, who found that the orbital speeds of stars around spiral galaxies were too fast to be accounted for by the gravitational attraction of the stars and gas in the galaxies. Even earlier, in the 1930s, Fritz Zwicky had argued that galaxies in dense clusters of galaxies were moving around too fast to be understood unless clusters were permeated by dark matter. To explain the origin of structure in the universe, the dark matter could not be in the form of protons and neutrons, which are collectively known as baryons, so it became known as *nonbaryonic dark matter*. One possibility was neutrinos, ghostly particles that interact very little with other matter but are known to be created in radioactive beta-decay. These pervade the universe in great numbers, left over from the hot phase of the Big Bang. They are normally considered to be massless and to move at the speed of light like photons, but if these particles had a small nonzero mass, they would be one option for the dark matter. Neutrinos with mass are called *hot dark matter*, and a universe in which a significant fraction of matter is in this form is expected to evolve in a very specific way. First, very large structures, clusters of galaxies, would form, and then later these would fragment into galaxies. By 1990, the consensus was that the universe could not have evolved in this way. Instead cosmologists preferred a 'bottom-up' picture, in which small structures form first and merge together to make galaxies and then galaxies aggregate together into clusters of galaxies. For such a picture to work, the dark matter must consist of particles that do not move around at speeds close to the speed of light like neutrinos, and this became known as *cold dark matter*.

The discovery of the cosmic microwave background radiation was one of the most significant scientific discoveries of the twentieth century and opened the way to understanding the origin of the universe itself. The hot Big Bang model, with the universe dominated by radiation for its first 30,000 years, became firmly established. But as interest shifted to how galaxies had formed, it became clear that the simplest possible models did not work. Some new kind of matter, cold dark matter, had to play a dominant role. With the launch of the *Infrared Astronomical Satellite* (*IRAS*) in 1983, infrared astronomers began to play a part in cosmology, explaining the origin of the Galaxy's motion through the cosmic frame and trying to measure how much dark matter there is in the universe.

7

The *Infrared Astronomical Satellite* and the Opening Up of Extragalactic Infrared Astronomy: Starbursts and Active Galactic Nuclei

The launch of the *Infrared Astronomical Satellite* (*IRAS*) on 25 January 1983 was a very exciting moment in the history of infrared astronomy. It had been designed to survey the whole sky at far-infrared wavelengths between 10 and 100 microns. When the helium coolant ran out ten months later on 22 November, *IRAS* had surveyed 96% of the sky. Over the next few years there were a string of discoveries that still comprise the core of our knowledge of the dusty universe: the zodiacal dust bands, the link between Apollo asteroids and comets, the infrared 'cirrus', debris disks and protoplanetary systems, ultraluminous and hyperluminous infrared galaxies, dust tori around active galactic nuclei, young stellar objects and 'protostars', and the origin of our Galaxy's motion through the cosmic frame.

THE *IRAS* STORY

During the 1970s, Dutch astronomers had been exploring the concept of a dedicated infrared satellite and managed to get U.S. astronomers and NASA interested in the idea, with the latter proposing that the wavelength range be extended to 100 microns. In 1975, Nancy Boggess, head of infrared astronomy at NASA, assembled a group of infrared astronomers at Snowmass, a ski resort in Colorado. To their surprise, she announced that NASA intended to build a space facility for infrared astronomy and to operate it from the instrument bay of

I have made extensive use of the reviews by Soifer et al. (1987) and Beichman (1987).

the space shuttle. They had even given it a name, the *Shuttle Infrared Telescope Facility*, or *SIRTF*. The astronomers argued instead for a smaller survey satellite in Earth orbit which would map the whole sky in a few weeks, a proposal that Neugebauer and others had originally made in 1971. The survey satellite concept did not make much progress until the Dutch began to propose a joint U.S.-Dutch mission. In 1976, on the initiative of the Dutch astronomers, the United Kingdom was also invited to join what became the *Infrared Astronomical Satellite*, or *IRAS*. Gerry Neugebauer led the U.S. team and was overall leader of the *IRAS* science team; the Dutch team was led initially by Reiner van Duinen and then by Harm Habing; and Dick Jennings of University College London led the U.K. team.[1] The United States provided the cryostat, telescope and survey detector arrays. The Dutch provided the spacecraft, a low-resolution spectrometer and a chopped photometric channel. The United Kingdom was to provide the ground station and planned to build a 200-micron photometer for *IRAS*, but after some development work this was abandoned because it was felt there was insufficient time to deliver it. It turned out that *IRAS* was delayed by several years, so there might well have been time.

The seven years of constructing and testing *IRAS*, and making preparations for its launch, seemed interminable, but in fact this was quite quick compared with several recent space astronomy missions. The *Spitzer* and *Herschel* missions both took more than 20 years from conception to launch. I remember 1976–83 as a seemingly endless series of meetings in windowless rooms. The *IRAS* team contained some powerful personalities who had played a pioneering role in the development of infrared astronomy. Gerry Neugebauer, who had carried out the Two Micron Survey, was a very focused and effective leader. Wiry, tense and with a colourful turn of phrase often peppered with expletives, he kept a diverse team on its toes. Frank Low, who had pioneered far-infrared and submillimetre astronomy, was a large man capable both of warmth and volcanic rage. There were some memorable explosions between these two giants of infrared astronomy. Both were key figures in the delivery of *IRAS*. The two other major figures in the development of the *IRAS* focal-plane instrument were Fred Gillett, the leading figure in 2–13-micron spectrophotometry, with a gentle and slightly dreamy manner, and Jim Houck, who had real experience launching infrared instruments into space and an engaging and open style. And behind the members of the *IRAS* Science Team was a vast army of engineers and scientists who designed and built the satellite and made it work.

The construction of *IRAS* required the solution of some novel technical problems. The cooling of the mirror and focal-plane instrumentation to a temperature just a few degrees above absolute zero was to be provided by superfluid helium. When helium is cooled very close to absolute zero, it becomes a 'superfluid', with essentially zero viscosity. Such a material had never been handled in the zero gravity of a space orbit before. There were great difficulties with the valve that released helium from its storage tank. This valve failed during testing, and there were anxious moments after launch regarding whether it would function correctly. There were 62 detectors in the focal plane, arranged in eight lines, two for each of the four wavelength bands centred on 12, 25, 60 and 100 microns. Each source on the sky crossed over at least two detectors in each band, providing important confirmation of the reality of the detected sources.[2] The shorter-wavelength doped-silicon detectors were based on existing military technology, but the longer-wavelength germanium-gallium detectors performed poorly and were rebuilt at the Jet Propusion Laboratory (JPL) under Jim Houck's direction, using material he had bought off the shelf for a few dollars. The detector preamplifiers were destroyed during testing, and the focal plane had to be rebuilt, also at JPL. New preamplifiers were designed and built at Frank Low's Tucson laboratory.

IRAS was eventually launched on a Delta rocket from Vandenburg Air Force Base in California on 25 January 1983 (Figures 7.1 and 7.2), a very memorable day for me. Most of the *IRAS* scientists stood on a grassy knoll about a mile from the launch. As night fell and the terminator that divides night from day crossed us, the sky lit up as the engines of the Delta rocket were ignited and *IRAS* rose slowly into the sky. There was a flash as the second stage ignited. For a second I thought *IRAS* had blown up. Gerry Neugebauer swore. But all was well, and *IRAS* was placed into a perfect orbit. There was jubilation that evening among the hundreds of scientists and engineers who had worked on *IRAS* for many years. *IRAS* was to continue its orbit along the day–night terminator, with its telescope always pointing away from Earth and 90 degrees away from the Sun, for the next ten months.

THE *IRAS* DATA CENTRES

There were two *IRAS* data centres, one at Rutherford Appleton Laboratory (RAL) at Chilton, United Kingdom, where the ground station and satellite control centre were based, and one at the Jet Propulsion Laboratory in Pasadena, California. The team at RAL, led by Peter Clegg,

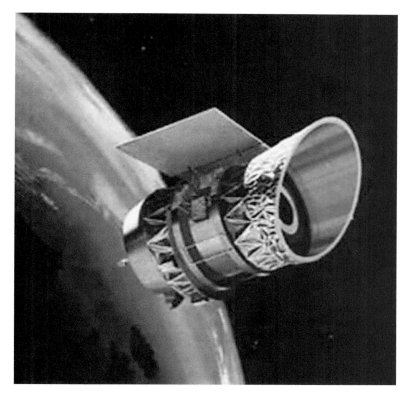

Figure 7.1. Artist's impression of the *IRAS* satellite in orbit.

had to plan the commands to the satellite each day and check out each day's data as soon as transmission to the ground station was complete. Most of the Dutch and U.K. scientists were based there, together with Fred Gillett and George Aumann from the United States. The team at JPL was to perform the final data analysis and prepare the catalogues and sky images from the all-sky survey in the four *IRAS* wave bands of 12, 25, 60 and 100 microns. The rest of the U.S. team was based here, together with several Europeans, including George Miley from the Netherlands and myself from the United Kingdom. The first weeks were very intense and dramatic. Immediately after launch there was a panic because each time *IRAS* passed over the ground station at RAL it was found to have overridden its commands and parked itself in safety mode. This was traced to a tiny error in the on-board software. A new version was quickly written and uplinked to the spacecraft computer. The plan had been to eject the cover from the helium dewar contain-ing the telescope a week after launch, but it became clear that the

Figure 7.2. The launch of *IRAS* on 25 January 1983.

cover was warmer than expected and was wasting helium coolant. It was decided to eject the cover straight away, and the *IRAS* telescope had its first glimpse of the sky. It was clear almost immediately that *IRAS* was a success and was working better in space than it had on the ground. The very first scan around the sky showed a strong peak at the Galactic plane and dozens of discrete sources of radiation all around the sky. After various tests and calibrations, the team at RAL launched *IRAS* into its 'Minisurvey', a planned quick survey of part of the sky that was to be used to test the quality of the main survey and to give us some science results if the satellite failed suddenly.

VEGA AND DUST DEBRIS DISKS

Different members of the *IRAS* team focused on different aspects of the data, and everywhere we looked discoveries were made. At RAL they were concentrating on calibrating the detectors by using asteroids and bright stars. One of the key calibration stars was Vega (Alpha Lyrae), the brightest star in the constellation of Lyra. Astronomers have chosen this as the prototype of the stellar magnitude system, and Vega is defined to have zero magnitude in all wave bands. The magnitude scale is logarithmic, and the magnitudes of other stars are then defined by how much brighter or dimmer they are than Vega. For example, if a star is 5th magnitude in a particular wave band, then it is a hundred times fainter than Vega.

However, Fred Gillett and George Aumann were having problems with Vega. It just did not give consistent results in the *IRAS* bands. They realized that it had a dust shell whose emission started to dominate over emission from the surface of the star at a wavelength of around 30 microns.[3] This was very surprising for a hot, luminous star like Vega. Up to this time, dust shells had been found only around cool red giant stars. This turned out to be a truly momentous discovery because it became clear that the disk of dust around Vega had not been ejected from the star as was the case for the red giant circumstellar dust shells, but was primordial material left over from the formation of the star. The Vega debris was analogous to the asteroid and Kuiper belts in the Solar System, and it was soon found that many other stars are also surrounded by such *debris disks*.

THE *IRAS* COMETS

Another exciting discovery by the team at RAL soon followed. Because each day's data were examined as soon as they were downloaded, there was the possibility of spotting new Solar System objects such as comets or asteroids, which would be characterized by their motion relative to the stars and galaxies. *Asteroids* are simply small Solar System objects that are not planets or moons, ranging in size from tens of metres to a thousand kilometres. The larger ones are solid objects, but some of the medium-sized asteroids are known to be loose aggregates, known as 'rubble piles', that result from a destructive collision with another asteroid. *Comets*, on the other hand, tend to be loose aggregates of gas, dust and ice and have been described as 'dirty snowballs'. Some, like Comet Halley, are known to have solid nuclei tens of kilometres across. As comets come near the Sun, they develop a glowing halo, the 'coma', and long tails of dust and gas, which are swept out of the comet by the pressure of the Sun's light and the solar wind. On the suggestion of Jack Meadows, a group had been set up to look through the *IRAS* data each day for moving objects. Early in the mission, they found a bright comet. The first orbit calculated for this comet showed it coming very close to Earth. Indeed a collision with Earth could not be ruled out. I remember the JPL-based science team crowding into Gerry Neugebauer's office for a telephone conference with our colleagues at RAL to decide what to do. Some of us argued that we should

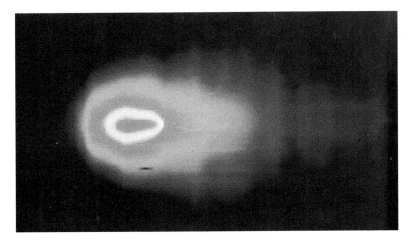

Figure 7.3. Comet IRAS–Araki–Alcock, as mapped by *IRAS* at 25 microns (the contours just denote different brightnesses). Its nucleus is 10 kilometres across, and it was the closest approach to Earth by a comet in 200 years, passing only 7 million kilometres away.

issue a public warning, on the grounds that if we did not and this comet did hit the Earth, we would be pilloried. After heated discussion, a scientist's typically cautious decision was taken to get more data and calculate the orbit more precisely. In the end, the comet did not hit the Earth, though it passed closer to Earth than any comet in the past 200 years and provided a splendid spectacle in the evening sky. The *IRAS* team failed to notify the Minor Planets Bureau about the comet and as a result two amateur comet-searchers, Genichi Araki and George Alcock, independently discovered it. The compromise was that the comet became known as Comet IRAS–Araki–Alcock (Figure 7.3).[4] If it had hit the Earth, there could have been a major catastrophe, because the nucleus turned out to be 10 kilometres across.

There were numerous other *IRAS* comets and asteroids found during the mission. One of these, Phaethon, demonstrated the link between comets and the Earth-orbit-crossing 'Apollo' asteroids.[5] While most asteroids orbit the Sun in a belt between Mars and Jupiter, a few are found in orbits that cross the Earth's orbit, and these are the Apollo asteroids. They are of great interest because occasionally one of them may hit the Earth, with potentially dramatic consequences. Phaethon (IRAS comet 1983TB) was discovered in the *IRAS* data by Simon Green and John Davies on 14 October 1983. It looked exactly like an Apollo asteroid, with no tail or coma, but it was found to be in an orbit identical to the Geminid meteor stream. Now meteor streams are known to be the debris of comets. The famous Perseid meteor showers in August result from the Earth encountering debris from Comet Swift–Tuttle, for example. The fact that Phaethon had the same orbit as the Gemini meteor stream demonstrated that Phaethon was an extinct comet nucleus and the source of the Geminid meteors. It no longer has the halo of ice, gas and dust responsible for the dramatic tails generated as comets approach the Sun. Phaethon has the distinction that it makes a closer approach to the Sun than any other known asteroid, with a closest distance of only 0.14 AU, less than half the radius of Mercury's orbit. The astronomical unit, or AU, is the mean distance of the Earth from the Sun, 150 million kilometres. At its farthest point from the Sun, Phaethon reaches 2.4 AU, in the zone of the asteroid belt between Mars and Jupiter, and it orbits the Sun every 524 days. Apollo asteroids are watched with special interest and concern in case they might collide with Earth. Phaethon is predicted to pass within three million kilometres of the Earth in 2093.

IRAS GALAXIES

At the JPL data processing centre, we were grappling with the massive data reduction code and trying to tune the many parameters in the code. My job was to look after the *IRAS* Point Source Catalog, working with Tom Chester and his team of software engineers. We spent a lot of time staring at plots of detector response versus time, noting what seemed to be sources and what seemed to be artefacts, and checking whether the code for source detection and confirmation processed them correctly. When we first came across the Large Magellanic Cloud in the data, a few weeks after launch, the detector scans seemed to go crazy and it looked as though the satellite had run into some major debris in its orbit. At the time we were not too sure exactly where on the sky the satellite was pointing. Gradually it dawned on us that the telescope had crossed a group of very bright sources in the Large Magellanic Cloud (Figure 7.4). From this it was possible to correct the *IRAS* pointing model – essentially an on-board clock was a few seconds off, so our position along the direction the satellite was travelling across the sky had been way off.

Every morning the team met to discuss progress, under the chairmanship of Tom Soifer. To improve our ability to extract astronomical sources from the *IRAS* data streams, we endlessly analyzed the data from the Minisurvey, changing the processing model a bit each

Figure 7.4. Final *IRAS* map of part of the Large Magellanic Cloud, the Galaxy's nearest neighbour. The Tarantula Nebula is the bright patch to the left.

time and seeing whether particular problem sources emerged from the processing more sensibly. Now that we had the correct source positions, we could compare the infrared sources with existing optical catalogues. Our first list of identified sources from the Minisurvey became the basis of several of the first *IRAS* papers. From the *IRAS* colours alone we could distinguish stars, galaxies and other objects such as planetary nebulae.[6] In the Galactic plane there were bright H II regions and small dust clouds like Barnard 5.[7] Some of the galaxies were surprisingly luminous.[8]

Not all the sources in the Minisurvey could be identified with known objects on optical photographs. Jim Houck led a study of nine *IRAS* sources from the Minisurvey which appeared optically very faint or, in the case of four of the sources, blank.[9] All but one turned out to be faint galaxies, luminous in the infrared; the ninth was a wispy piece of 'cirrus' emission from interstellar dust in the Galaxy. Several of us, led by Tom Soifer, worked on a particularly spectacular luminous

Figure 7.5. A *Hubble Space Telescope* image of the ultraluminous infrared galaxy Arp 220. There are two nuclei, relics of the two galaxies that have merged together.

infrared galaxy, the peculiar galaxy Arp 220, (Figure 7.5) from Halton Arp's *Catalogue of Peculiar and Interacting Galaxies*. This had an infrared luminosity greater than a million million (10^{12}) solar luminosities and became the prototype of the class of ultraluminous infrared galaxies.[10] It was clear that the *IRAS* survey was finding thousands of distant and very luminous starburst galaxies. Later follow-up would show that many of them are interacting or merging galaxies, with plumes and very disturbed appearances. The picture took shape that these very luminous starbursts are triggered by major mergers between galaxies. The more spectacular of these mergers were nicknamed 'train wrecks' because the two merging galaxies seemed to have wrecked each other.

With his background in radio astronomy and the study of radio galaxies and quasars, George Miley focused on the *IRAS* detections of active galaxies. He led a study of the radio galaxy 3C 390.3,[11] which had surprisingly strong emission at 25 microns, suggestive of warm (180 K) dust. This led to the realization that the central massive black holes of active galactic nuclei (AGNs) are surrounded by a doughnut-shaped dust torus, which intercepts some of the visible and ultraviolet light from hot gas swirling around the black holes (Plate IV). Active galactic nuclei look different depending on whether they are seen through this dust torus or face-on.[12] From face-on we can see the central stellar source clearly at all wavelengths, but from edge-on much of the optical and ultraviolet light is extinguished by dust, though we can still see the infrared and radio output.

George Helou, Tom Soifer and I found a very interesting correlation between far-infrared luminosity and radio luminosity in star-forming galaxies.[13] At first we thought this might be caused by radio emission from supernova remnants, which are the relics of the death of massive stars and which are strong radio sources, but we realized the radio emission of star-forming galaxies was too strong to be explained this way. A profusion of electrons must be accelerated close to the speed of light within the supernova remnants. These fast-moving electrons must then leak out of the supernova remnants and generate radio emission as they spiral through the galaxy's general magnetic field.

THE INFRARED 'CIRRUS' AND THE ZODIACAL DUST BANDS

The thing that really excited Frank Low about *IRAS* was its ability to map large areas of the sky, and he posted long strip maps on the walls of the JPL corridors. These showed two exciting new phenomena: extended emission from interstellar dust, the infrared 'cirrus', and structure in the emission from the interplanetary dust in our solar system, the zodiacal dust bands. Before launch we had been aware of Allan Sandage's 1975 work on the cirrus-like distribution of scattered optical light from interstellar dust at high latitudes, and we tried to estimate whether we might detect this in the far infrared. Still, the ubiquity and brightness of the infrared 'cirrus' took us all by surprise. Interstellar dust had been known from its dimming and reddening of visible light since Trumpler's work in 1930, but here for the first time we could see the dust clouds directly radiating at us (Figure 7.6). If we could look up at the night sky with infrared eyes, we would see these shining clouds and filaments of dust across almost the whole sky.

The zodiacal dust bands were completely unexpected. Four bands were clearly visible parallel to the plane of the Earth's orbit

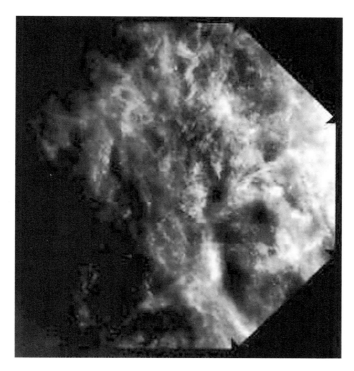

Figure 7.6. Infrared 'cirrus', emission from clouds of interstellar dust, at the South Ecliptic Pole.

around the Sun (the *ecliptic plane*), and the displacement of these bands from the plane was the clue to their nature. They were in fact debris from collisions between members of asteroid families. An *asteroid family* is a group of asteroids within the main asteroid belt between Mars and Jupiter which have similar orbits. They were discovered in 1918 by the Japanese astronomer Kiyotsugu Hiroyama (1874–1943).[14] A collision between two members of an asteroid family generates a cloud of dust that is gradually smeared out along the orbit of the whole asteroid family to produce the *IRAS* dust bands.

PLANET X?

While we were working on the new *IRAS* results in the winter of 1983, a rumour emerged that we had discovered a tenth planet in the solar system. This rumour was published in the *Washington Post* on 30 December 1983. A tenth planet, Planet X, had been postulated for some time because Neptune's orbit has unexplained anomalies that could be caused by a planet out beyond Neptune and Pluto. In fact, when Clyde Tombaugh discovered Pluto in 1930, he was looking for the planet that was the cause of these anomalies. Neptune itself had been discovered from anomalies in the orbit of Uranus. Pluto turned out to be far too small to explain the anomalies in Neptune's orbit, so Tombaugh's discovery was fortuitous. Ironically, *Voyager 2*'s flyby of Neptune in 1989 eliminated the anomalies by showing that Neptune's mass needed to be slightly revised.

If Planet X existed, then *IRAS* would have had a good chance of finding it, and several of us on the *IRAS* science team had this in the back of our mind as we scanned the *IRAS* data. One day in December 1983 a group of us were looking on a computer screen at a new map of the central regions of the Milky Way that had been assembled from the *IRAS* data. Someone pointed out a bright source near the centre of the Galaxy. Its position turned out to lie near the ecliptic plane. Could this be the mysterious Planet X? We collected all the observations of the object from the *IRAS* archive. It was indeed a cool source, but the observations taken at different times all fell on the same position on the sky, whereas Planet X would have been moving slowly across the sky. We found that a spectrum of the object had been taken by the Dutch spectrograph on *IRAS*, and this showed deep silicate absorption. This suggested an opaque dust cloud surrounding a red giant star. It was definitely not Planet X, and we knew this within a few days of first noticing the object.[15]

Unfortunately someone got hold of the wrong end of the stick and reported at a meeting at NASA Headquarters that we might have discovered Planet X. Someone else in that meeting phoned the *Washington Post*, and they ran the story on 30 December 1983:

> At Solar System's Edge Giant Object is a Mystery— A heavenly body possibly as large as the giant planet Jupiter and possibly so close to Earth that it would be part of this solar system has been found in the direction of the Constellation Orion by an orbiting telescope called the IRAS. So mysterious is the object that astronomers do not know if it is a planet, a giant comet, a 'protostar' that never got hot enough to become a star, a distant galaxy so young that it is still in the process of forming its first stars, or a galaxy so shrouded in dust that none of the light cast by its stars ever gets through. "All I can tell you is that we don't know what it is," said Gerry Neugebauer, chief *IRAS* scientist.

There was some confusion here between the 'Planet X' investigation and the blank-field sources that Jim Houck had been studying and which were part of the press conference put out about the first batch of science results. I got a phone call from a science journalist at *New Scientist* while they were preparing to report this story, and I told him that our object was definitely not a planet. They ran the story anyway. Amazingly this rumour has an Internet life of its own: how *IRAS* discovered Planet Nibiru and NASA then covered it up. Tom Chester has a Web page on this story,[16] which includes this wonderfully nonsensical synopsis of the Planet Niburu theory:

> Sirius and the Sun are gravitationally tied systems, orbiting Alcyone in 'the Pleiades Quadrant'. 500,000 years ago another planet in the Sirius system, Nibiru, 'strayed off course' and 'drifted our way', was 'unwittingly captured by our Sun' and is now at the edge of our Solar System, inhabited by 'a reptilian super-race'. To stay warm, they nabbed some gold from Earth, via a spaceport in Kuwait, and now have a nice gold heat-shield all around their planet. Later a collision of one of Nibiru's moons with Earth created the Pacific Ocean, with all the ejecta creating the asteroid belt.

A few months later, I conducted a more systematic search for Planet X in the *IRAS* data, hunting through the database of sources that had been seen only on one orbit but not confirmed on the next orbit. During the first six-month survey, a source would typically be observed on two or three successive orbits of the spacecraft. A moving object would appear in this database as a series of entries slightly displaced from each other. I did find a really excellent candidate with three detections displaced by a few arcminutes and the spectrum of a 100 K blackbody. For a few days, I thought I really had found a new planet in the Solar System. Unfortunately, when our resident Solar System expert, Russ Walker, checked it against the minor planet and comet database, it turned out to be a comet that was already known, Comet Bowell.[17] This was a comet that had not come into the inner Solar System but had reached its minimum distance from the Sun at about the distance of Jupiter, at which point it had been found by Edward Bowell. When I found it in the *IRAS* data it was on its way back to the outer reaches of the Solar System.

GROUND-BASED FOLLOW-UP OF THE *IRAS* SURVEY

When the first survey of the sky had been completed after six months, there was a debate amongst the science team about how the remaining time on the satellite should be used. The helium coolant might last for as much as another six months. Some of us were keen for a second sky survey to be carried out to improve the reliability of our source detection and sensitivity. Frank Low did not, at the time, have a very high opinion of the all-sky survey and argued forcefully that the time should be used for pointed observations. Luckily Gerry Neugebauer and Tom Soifer backed the survey idea. Because the Minisurvey had been carried out on a strip rather close to the Milky Way, which passed through the Orion, Taurus and Ophiuchus star-forming regions, clear detection of galaxies had seemed quite problematical. By July 1983, when we started to scan in regions of the sky well away from the Milky Way, free of the star-forming regions and dense dust clouds in the Galaxy, the infrared sky looked very different and we started to see thousands of galaxies, especially at a wavelength of 60 microns.

It was clear that *IRAS* was going to provide a wonderful all-sky sample of galaxies (see Plate V). To learn more about these galaxies, we needed observations from ground-based telescopes to assess the morphology of the galaxies and measure their redshifts. Tom Soifer and I agreed to lead two different ground-based surveys. He led a Caltech team that focused on a sample of bright, nearby *IRAS* galaxies,[18] while I set up a U.K. collaboration to study a complete sample of *IRAS* galaxies at the North Celestial Pole, including much fainter *IRAS* sources. The Caltech survey emphasized the role of star formation in luminous infrared galaxies. Their sample was to form the basis of many subsequent detailed studies with ground-based telescopes and with space missions such as *ISO* and *Spitzer*. Our north polar survey confirmed that most of the 60-micron sources in the *IRAS* Point Source Catalog away from the Milky Way are indeed galaxies.[19] The typical depth of the *IRAS* galaxy survey turned out to be about 500 million light years. An important product from both surveys was the *luminosity function* of *IRAS* galaxies, which gives the probability that a galaxy will have a particular infrared luminosity. We were able to estimate this for galaxies with infrared luminosities ranging from one-hundredth to one hundred times that of our own Galaxy. Much interest focused on which types of galaxies were being detected as prominent infrared sources. Edwin Hubble had classified galaxies into *spirals*, which show spiral arms, clumps of newly formed massive stars and an overall blue colour, and *ellipticals*, which have a smooth elliptical shape and are composed of old red stars. Most *IRAS* galaxies were spiral galaxies undergoing strong star formation, and their optical spectra showed bright emission lines associated with the formation of young, massive stars. Elliptical galaxies, which appear to have completed their star formation long ago, tended to be rather weak infrared sources.[20]

These studies of the optical appearance of infrared galaxies showed that some, especially the more luminous ones, looked quite peculiar. As we look towards higher infrared luminosities, we see an increasing incidence of galaxy interactions and mergers.[21] The most luminous infrared galaxies seemed to be 'train wrecks', where two large spiral galaxies had crashed into each other, severely disrupting both galaxies. In a picture developed by Dave Sanders, the endpoint of such a merger would be an elliptical galaxy, often with a bright quasar in its nucleus.[22] The idea is that the gas in the merging galaxies is compressed to generate a huge starburst. As the gas is exhausted and mainly driven off by supernova explosions, the more massive blue stars die, leaving only lower-mass red stars shining. The galaxy becomes a red, elliptical galaxy, relatively free of gas and dust. Some residual gas from the merger event finds its way to the nucleus of the galaxy. There it is accreted onto a black hole in the nucleus, getting heated to very high temperatures to give the very bright quasar phase. This picture provided a direct connection between the growth of a massive black hole, being fed by gas from the merger event, and the growth of the central stellar bulge of a galaxy as a result of the star formation triggered by the merger.

The sources about which we could say most about their infared emission were those detected in all four *IRAS* bands at 12, 25, 60 and 100 microns. George Helou showed that star-forming galaxies occupied a sequence of far-infrared colours that could be interpreted as a mixture of two components: emission from interstellar dust in the galaxy absorbing the general starlight of the galaxy and bursts of star formation in dense molecular clouds similar to that seen in the archetypal nearby starburst galaxy M82.[23] My postgraduate John Crawford and I selected over one thousand *IRAS* galaxies detected in all four bands and set out to model their 12–100-micron spectra in detail.[24] We saw that the infrared spectra of most galaxies could be thought of as a mixture of three simple components: the two components proposed by Helou and, in addition, mid-infrared emission from a dust torus surrounding a quasar in the nucleus of the galaxy.[25]

COSMOLOGY WITH *IRAS*

The *IRAS* galaxy survey offered intriguing cosmological possibilities. This was the first time we had an all-sky survey of galaxies covering both Northern and Southern Hemispheres in a homogeneous way. Working in the infrared allowed us to probe closer in towards the centre of our own Galaxy without worrying about the effects of extinction by interstellar dust. The exciting prospect was to map the three-dimensional galaxy distribution in a substantial volume of the local universe, out to 500 million light years, and study the origin of the Galaxy's motion through the cosmic frame, as evidenced by the cosmic microwave background dipole (see Chapter 6). Groups in the United States and United Kingdom embarked on ambitious redshift surveys of *IRAS* galaxies to make such a 3-D map a reality.

Andy Lawrence and I teamed up with several leading U.K. cosmologists to carry out a redshift survey over the whole sky. The plan was to measure 2000 galaxy redshifts sampled randomly from the whole *IRAS* survey. By good fortune, the 4.2-metre William Herschel Telescope was just completing 'commissioning', or final testing, on La Palma. During the commissioning process, we were successful in bidding for one month of time on the telescope, in December 1987, on a 'shared risks' basis. If the telescope was working, we could use it, but if it needed adjustment or repair, we had to hand it over to the engineers. In fact, the telescope worked perfectly and we were able to measure the spectra of most of the Northern Hemisphere galaxies in our survey. A highlight was the measurement of over one hundred galaxy redshifts in a single night. At that time, spectroscopy was a laborious procedure, observing one galaxy at a time, so this seemed like a tremendous achievement. Nowadays there are multiobject spectrographs that measure hundreds of objects simultaneously. The Southern Hemisphere objects in our survey were measured during a couple of observing runs at the Anglo-Australian Telescope at Siding Spring.

In parallel with our all-sky survey, Mark Davis of Berkeley was leading a redshift survey of over 2600 brighter *IRAS* 60-micron galaxies, later extended to over 5000 galaxies. Finally, Will Saunders led a survey which completed redshift measurements of all the *IRAS* galaxies on the sky to the effective limit of the *IRAS* 60-micron survey, a total of over 15,000 galaxies. From these surveys, we could map the density distribution of galaxies in the local universe in an unbiased way (Figure 7.7). The clusters of galaxies and the voids between them could be clearly seen, traced out in the *IRAS* galaxy distribution. Each of these clusters is exerting a small gravitational pull on our Galaxy. We could now combine all these pulls together and see whether the net attraction of galaxies in the local universe could explain our Galaxy's motion with respect to the microwave background. The major result from these surveys was to demonstrate that the origin of our Galaxy's motion with respect to the cosmic microwave background was the net gravitational attraction of galaxies and clusters of galaxies within 500 million light years.[26] We had shown why our Galaxy is moving through the cosmic frame of reference from which everything would look isotropic. This was a major cosmological discovery of *IRAS*.

We could also try to estimate the average density of matter in the universe to account for the observed speed of our Galaxy's motion through the cosmic frame. The density of the universe is an important quantity because if it is high the self-gravity of the universe will halt the expansion of the universe and everything will eventually collapse together into a Big Crunch. If the density is low, the expansion of the universe will continue forever. In between there is a *critical density* at which the self-gravity of the universe is just strong enough to keep on slowing the expansion forever without actually turning the expansion into a contraction. Because in Einstein's General Theory of Relativity there is an intimate connection between the matter in the universe and the spatial geometry, a universe with the critical density is also spatially flat. If the density is higher or lower than the critical value, the geometry of the universe is curved.[27]

The first attempt to measure the density of the universe using the *IRAS* survey was made in 1986 by Amos Yahil from Stony Brook, New York, my researcher David Walker and myself.[28] Amos visited us at Queen Mary College and pointed out that we did not need to know the redshifts of the galaxies in a survey to estimate the mean density of the universe; we just needed the average distribution of redshifts, and we had just determined this from our north polar survey. We found a rather high mean density of the universe, and at first this was confirmed by our all-sky redshift surveys. Ninety per cent of the matter would be cold dark matter and just 10% ordinary (baryonic) matter. This meant that the total density of the universe seemed to be close to the critical value.

Theoretical cosmologists quite liked this critical density model of the universe because of its simplicity and also because a spatially flat geometry was a prediction of a new model for the early universe, *inflation*. In 1980, Alan Guth had proposed that in its very early stages the universe went through a phase of very rapidly accelerating expansion. His idea was that this was driven by a huge and transient 'vacuum energy density', which could be left behind by some major change in the state of the universe. We normally think of the vacuum as empty, but particle phyicists imagine it teeming with transient particles being formed and then annihilating. So the vacuum can in principle have a very high energy density, and this acts like a repulsive force, accelerating the expansion of the universe. This very brief period of inflation solved a number of puzzles about the universe, including the fact that when we look at the cosmic microwave background in opposite directions on the sky, we see exactly the same intensity of light, even though those regions have never been in causal contact with each other (the 'horizon' problem). In the inflationary model, the whole observable universe today would have been so tiny prior to inflation that there could have been causal contact between all parts of it.

One of the problems with our density estimate was that we had to make a correction for the areas of sky in which we could not measure galaxy redshifts because they were behind the confused regions of the central Milky Way. As we did larger redshift surveys and also started to model more carefully the effect of the missing areas, we found that our estimates of the mean density of matter in the universe were reduced well below the critical value. We shall see (in the next chapter) that the preferred cosmological model today following the analysis of data from the *WMAP* satellite is indeed spatially flat, but the density of matter is only 20% of the critical value and requires the additional ingredient of the mysterious 'dark energy'.

When we had the data from our first all-sky survey in hand, Will Saunders made three-dimensional maps of the density distribution of galaxies, and these were published in rather gorgeous colour versions in the science journal *Nature*.[29] More importantly, Will analyzed the statistics of the density variations, from the peaks in dense clusters to some dramatic voids. These results could be compared with the predictions from different cosmological models. The sensational result was that the distribution was inconsistent with the consensus model of the day, the 'standard' cold dark matter model. Basically there was more structure in the galaxy distribution on the largest scales than predicted by the model. We'll see in the next chapter what causes this excess structure on large scales.

There was also great interest in whether we could see evidence for evolution of the infrared galaxy population, changes with cosmic time in the average infrared properties of galaxies. As part of his PhD thesis with Jim Houck, Perry Hacking analyzed *IRAS* data at the North Ecliptic Pole, an area of the sky scanned every orbit by the *IRAS* satellite, and concluded that there were more faint sources than would be expected if there were no change of the infrared galaxy population with redshift.[30]

Figure 7.7. Three-dimensional map of the galaxy distribution, derived from the 'PSCz' all-sky *IRAS* galaxy redshift survey.

HYPERLUMINOUS INFRARED GALAXIES

While the all-sky *IRAS* redshift survey was continuing, I also set up a new collaboration with Carol Lonsdale and others at Caltech to carry out a redshift survey with the William Herschel Telescope of fainter *IRAS* galaxies in a 1400 square degree area of the sky which was particularly free of cirrus emission. Caltech had constructed a new *IRAS* Faint Source Survey (FSS) Catalog, which was more than a factor of two deeper than the *IRAS* Point Source Catalog. The latter had been constructed in a rather conservative way to ensure the reliability of the sources. Our FSS galaxy redshift survey, the data reduction for which was carried out by my postgraduate student Seb Oliver, yielded a startling discovery, the prototype hyperluminous infrared galaxy F10214+4724, a redshift 2.3 galaxy with a luminosity in excess of 10^{15} solar luminosities (one thousand million million times the luminosity of the Sun, and one hundred thousand times the luminosity of our Galaxy). Since most *IRAS* galaxies had redshifts < 0.3, this galaxy was quite a surprise. The spectrum, with several strong emission lines, took a while to understand. Eventually Andy Lawrence recognized the similarity of the spectrum with some known radio galaxies and deduced the redshift, but it still took us some further puzzling to work out which was the optical counterpart of the infrared source. As we were writing up the results for *Nature*, the extraordinary luminosity of the object dawned on us. At the time it was the most luminous object known in the universe.[31]

As soon as we had announced the object, there was a rush to observe it with submillimetre and microwave telescopes. We soon had detections at several submillimetre wavelengths, and line emission from carbon monoxide was also detected from F10214+4724.[32] The mass of molecular gas was estimated to be a remarkable one hundred billion (10^{11}) solar masses. This was very exciting for molecular-line astronomers, opening up the prospect of a cosmological role for their field. It has, however, proved to be quite hard work, and to date carbon monoxide emission has been detected from only 36 high-redshift galaxies.[33] Only one more of these was an *IRAS* source, a gravitationally lensed system with four multiple images known as the Cloverleaf because of its appearance. *Gravitational lensing* is a phenomenon predicted by Einstein's General Theory of Relativity where the bending of light from a distant galaxy around a foreground galaxy or cluster of galaxies can result in a highly distorted and magnified image of the background galaxy. Typically there are multiple images and curved arcs or a ring (see Plate VI). The galaxy F10214+4724 also turned out to be gravitationally lensed, and this accounted for a factor of about ten in its luminosity. Even correcting for this magnification, it was still a remarkably luminous infrared source. It is clearly a composite system, with a quasar-like nucleus responsible for the narrow emission lines and its mid-infrared emission, and an extraordinarily powerful starburst responsible for the emission observed at the longer submillimetre wavelengths.

Over the next few years, a number of other exceptionally luminous infrared galaxies were discovered, including the redshift 0.93 *IRAS* galaxy F15307+3252, for which Roc Cutri and his collaborators coined the term *hyperluminous* infrared galaxy.[34] Some of these were, like F10214+4724 and the Cloverleaf, gravitationally lensed, but imaging with the *Hubble Space Telescope* showed that many were not. Some were shown to contain huge reservoirs of molecular gas.[35] Even after correcting for the effects of lensing, the enormous far-infrared luminosities implied star-formation rates of over one thousand solar masses per year, compared with one solar mass a year in the Galaxy. These gal-axies appear to be consuming their gas a thousand times faster than our own. A typical large galaxy's mass of stars ($\sim 10^{11}$ solar masses) would be formed in only a hundred million years. Even if these bursts do not represent the very first star formation in the galaxies, they do represent their main episode of manufacturing heavy elements.

THEORETICAL MODELS FOR INFRARED SOURCES

The profusion of new results from *IRAS* gave new impetus to calcula-tions of theoretical models for infrared sources. The models of the early 1980s could be improved both by better models for interstel-lar grains and by dropping the assumption that the dust clouds were spherically symmetric. We already knew that there were several astro-physical situations where the latter would be a poor assumption.

By 1990, my postgraduate student Andreas Efstathiou had cracked the problem of analyzing the flow of optical and infrared radiation through a flattened dust cloud,[36] which allowed us to tackle problems with a disklike geometry. We applied this to models for newly form-ing stars and to dust tori around active galactic nuclei (AGNs).[37] For the latter we were able to explain the surprising absence of strong silicate emission features in AGNs by invoking a particular 'tapered disk' dust geometry. Edward Pier and Julian Krolik independently developed a code to study disklike geometries and applied it to AGN dust tori.[38] They demonstrated the importance of radiation pressure in support-ing and thickening the torus near its inner edge.

In 1998, Laura Silva and her collaborators developed models for the infrared emission of galaxies in which they considered the evolu-tion of gas, dust and chemical abundance and included both starburst and 'cirrus' components, and these have been widely used in galaxy evolution analyses.[39] In 2000, Andreas Efstathiou and his colleagues put together a more detailed model for starburst galaxies in which the whole evolution of the starburst – from formation of the massive stars to development of the expanding region of ionized hydrogen, the super-nova explosions when the massive stars end their lives, and the final dissipation of the outflowing shell of dust and gas – was followed.[40]

IRAS VIEW OF THE GALAXY

So far I have emphasized the *IRAS* point sources, but on large scales the infrared emission is mainly from zodiacal dust and from the Milky

Way. Which of these foregrounds is dominant varies with wavelength. At 12 and 25 microns, the zodiacal emission dominates, but the latter is quite hard to detect at 100 microns. The Milky Way dominates at 100 microns and is strong at 60 microns. The 100-micron emission from interstellar dust covers much of the sky, and its patchy 'cirrus-like' distribution sometimes makes it hard to detect point sources such as galaxies. As Chas Beichman puts it in the opening of his review on 'The *IRAS* View of the Galaxy and the Solar System': 'The Infrared Astronomical Satellite (IRAS) mapped the sky at 12, 25, 60 and 100 microns for 300 days starting 25 January 1982 and forever changed our view of the sky'.[41]

The diffuse emission at 60 and 100 microns comes mainly from interstellar dust grains with typical size 0.1 microns, while much of the diffuse Galactic emission at 12 and 25 microns is caused by transient emission from the very small grains, or large molecules, of polycyclic aromatic hydrocarbons (PAHs) discussed in Chapter 4. Away from the Milky Way, the 100-micron *IRAS* maps track the interstellar dust, which resides mostly in clouds of atomic hydrogen gas. This gas had already been mapped by radio astronomers using the 21-cm radio line of neutral atomic hydrogen.

In the Galactic plane itself, the *IRAS* extended emission maps traced out the giant molecular clouds where massive stars are forming and heating up the dust from their nascent cloud. Three very nearby molecular clouds, in Orion, Taurus and Ophiuchus, appear to lie slightly outside the great circle of the Galactic plane itself because they have a small vertical displacement from the midplane. One of the most dramatic images from *IRAS* was a map of the whole Orion constellation showing the dense molecular clouds filling the whole constellation (Figure 1.4 and Plate I). The familiar constellation of Orion the Hunter becomes completely unrecognizable, and it is almost shocking to realize how little of the true night sky our infrared-blind eyes are appreciating. For a moment, *IRAS* became our eyes.

Although 99% of 60-micron *IRAS* sources at high Galactic latitude are galaxies, the situation is very different at lower latitudes and at the shorter wavelengths of 12 and 25 microns. Most of the 250,000 sources in the *IRAS* Point Source Catalog are stars, seen either because of emission from their surfaces or through emission from surrounding circumstellar dust shells. Harm Habing made a very nice image of the Galaxy by plotting the distribution on the sky of stars with circumstellar dust shells detected at 12 and 25 microns (Figure 7.8).[42] For the first time, we see the central bulge and stellar disk of the Galaxy,

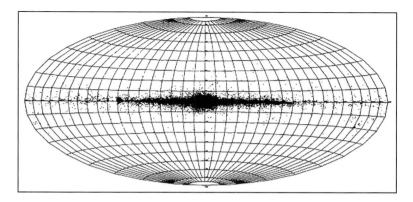

Figure 7.8. Map of *IRAS* stars with circumstellar dust shells on the sky, showing the disk and bulge of the Galaxy (Habing et al., 1985).

without extinction and without confusion from clouds of gas and dust. We can really feel ourselves to be on the edge of a disk galaxy, looking in towards the centre. The other ingredients of the Galaxy not seen in this image are the thin disk of gas and dust, the central nucleus of the Galaxy containing dense gas clouds, star clusters and a two million solar mass black hole, and the extended halo populated by old globular star clusters.

Most of the stars detected by *IRAS* are red giant stars which, because they have cool surface temperatures of 2000–3000 K, are much more prominent in the infrared than hotter stars. Some *IRAS* stars, like those plotted in Figure 7.8, clearly had prominent circumstellar dust shells, and the *IRAS* 12–100-micron brightnesses gave us a better idea about the cooler dust in these shells.[43] Several of the stars showed excess emission at 100 microns, hinting that the expanding dust shell has run into the interstellar medium and formed a dense rim of material.

Perhaps the most exciting discovery in Galactic astronomy made by *IRAS* was very young, low-mass stars and *protostars*, stars in the process of formation. One of the first of these protostars to show up in the *IRAS* data was the dust globule Barnard 5, which was detected in the *IRAS* Minisurvey. In 1986, Chas Beichman and his colleagues surveyed cold condensations of molecular gas in the Orion and Taurus clouds and found two types of objects: warm sources with broad energy distributions between 1 and 100 microns, almost always associated with T Tauri stars, and colder sources with spectra strongly peaked towards longer wavelengths, with no optical counterparts.[44]

While the T Tauri stars are believed to be young, low-mass stars which may have already started to burn hydrogen in their cores, the unidentified, colder sources appear to be younger objects with outflowing winds, surrounded by dense disks of dust and gas, and in some cases still accreting new material as they form. Both classes of objects have become known as young stellar objects (YSOs) and have been the subject of intensive study for the past 20 years, with special progress from the *Infared Space Observatory* (Chapter 10) and expected from the *Herschel Space Observatory* (Chapter 12). In a widely used classification of YSOs,[45] Class I corresponds to deeply embedded objects with heavily veiled spectra and associated outflows, Class II are classical T Tauri stars with ongoing accretion and substantial circumstellar disks, and Class III are fully formed stars whose light is partially extinguished and reddened by dust. Later data showed the need for an even more extreme, younger class of YSOs, Class O, where the extinction is so great that only submillimetre and far-infrared radiation are seen. These correspond to the true protostars, where the collapsing material building up a star has not yet become hot enough to ignite nuclear burning.[46]

IRAS was one of the most successful astronomical satellites ever launched and totally transformed our understanding of the cool universe. *IRAS* discoveries include the zodiacal dust bands, the link between Apollo asteroids and comets, the infrared 'cirrus', debris disks and protoplanetary systems, ultraluminous and hyperluminous infrared galaxies, dust tori around active galactic nuclei, young stellar objects and 'protostars', and the origin of the Galaxy's motion through the cosmic frame. We also tried to measure the mean density of the universe with *IRAS* and determine what kind of universe we inhabit. But to fully understand our universe we needed two new missions devoted to cosmology, *COBE* and *WMAP*.

8

The *Cosmic Background Explorer* and the Ripples, the *Wilkinson Microwave Anisotropy Probe* and Dark Energy

The *IRAS* surveys had demonstrated the origin of the dipole anisotropy in the cosmic microwave background (CMB) and seemed to have shown that the mean density of the universe could be close to the critical value. They had also shown that there was more structure on large scales than expected in the standard 'cold dark matter' model for the universe. In this chapter, I describe how the *Cosmic Background Explorer (COBE)* mission, launched in November 1989, measured the spectrum of the background radiation with unbelievable accuracy and then went on to detect the expected small fluctuations in the background that would be the precursors of galaxies and clusters of galaxies today. However, these fluctuations turned out to be stronger than expected, confirming the *IRAS* conclusion that an additional ingredient was needed to explain this structure. The choice seemed to be between massive neutrinos or a cosmological constant. This and many other conundrums about the universe were settled by the *WMAP* mission, launched in 2001. The beautiful detail of *WMAP*'s measurements of CMB fluctuations on both small and large angular scales heralded a new era of precision cosmology.

THE *COSMIC BACKGROUND EXPLORER* AND THE SPECTRUM OF THE COSMIC MICROWAVE BACKGROUND

During the *IRAS* mission of 1983, I recall that Mike Hauser, who took a lead role in developing the *IRAS* extended emission maps, was already working on the *Cosmic Background Explorer* mission. As far back as 1974, NASA had received three proposals for cosmological background radiation missions in response to a call for small- or medium-sized astronomical missions, which were known as the 'Explorer' series of missions. Although the mission selected on that occasion was *IRAS*,

NASA chose members from each of the three cosmic background proposal teams to get together and propose a joint satellite concept. In 1977, this team converged on the idea of a polar orbiting satellite, COBE, that could be launched by either a Delta rocket or the space shuttle. It would carry three instruments: a Differential Microwave Radiometer (DMR) to map anisotropies in the background radiation; a Far-Infrared Spectrometer (FIRAS) to measure the spectrum of the microwave background radiation; and a Diffuse Infrared Background Experiment (DIRBE) to map both near- and far-infrared radiation from the Galaxy and the background radiation from distant galaxies. NASA accepted the proposal on the condition that the costs be kept under $30 million, excluding the launcher and the costs of data and science analysis.

As a result of cost overruns in the Explorer program caused by IRAS, work on constructing the satellite at Goddard Space Flight Center (GSFC) did not begin until 1981. To save costs, the liquid-helium dewar on COBE and some of the detectors would be similar to those used on IRAS. The original plan to launch COBE on a space shuttle mission was shattered by the Challenger disaster of 1986. Eventually, a redesigned COBE was placed into a low Earth orbit, similar to that of IRAS, on 18 November 1989 aboard a Delta rocket.

The first nine minutes of observation with the FIRAS instrument already yielded a remarkable spectrum of the microwave background radiation, with a perfect blackbody spectrum. In fact, the theoretical curve passes through the data points so perfectly that people do not always realize that both observational data and the theoretical curve are being shown in this plot (Figure 8.1). This was the moment when cosmology ceased being a speculative piece of metaphysics and became a precision science. When John Mather showed this spectrum at a meeting of the American Astronomical Society on 13 January 1990 in Crystal City, Virginia, the huge audience of more than a thousand astronomers burst into spontaneous applause. As Mather described it:

> I took to the podium. After describing the instrument's principle of operation, I displayed a graph of the spectrum of the cosmic background radiation as revealed by FIRAS. 'Here is our spectrum,' I said. 'The little boxes are the points we measured and here is the black body curve going through them. As you can see, all our points lie on the curve.' The theoretical blackbody curve predicted how the blackbody radiation should look if it truly originated in the Big Bang.

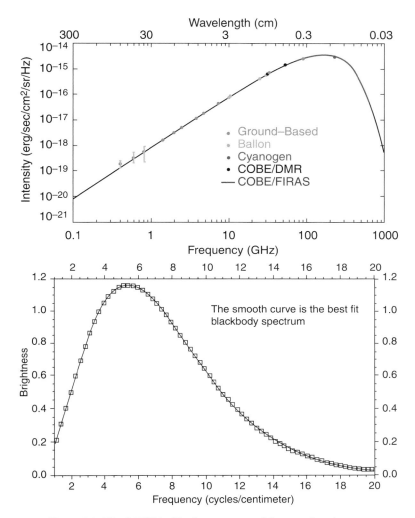

Figure 8.1. The 2.7 K blackbody spectrum of the cosmic microwave background measured by the *COBE* satellite (lower plot). The upper plot shows results from other experiments.

There was a moment of silence as the other scientists grasped the meaning of the data curve. Then the audience rose, breaking into spontaneous ovation. Blushing and with perspiration rising on my scalp, I stood there speechless before the huge crowd. It had never occurred to me that so many scientists would be there or that they would think the preliminary FIRAS result was so important.... Such displays of enthusiasm are rare at scientific meetings. I was entirely unprepared for it. Chuck Bennett told me later that he had never seen

> anything like it, before or since.... The FIRAS blackbody spectrum line, perfect in its harmony with theoretical speculation, confirmed what almost everybody there hoped and believed was true: that the Big Bang theory really did explain how the universe began.[1]

There was absolutely no sign of the possible deviation from a black-body spectrum that had been claimed by Paul Richards's group at Berkeley in 1979. A joint rocket experiment by the Berkeley group and researchers in Nagoya, Japan, in 1988 had seemed to further confirm the distortion. A deviation from a Planck blackbody spectrum would have been hard to understand in the context of the Big Bang model, so the *COBE* spectrum came as a great relief.

COBE AND THE RIPPLES

The FIRAS instrument on *COBE* had yielded its sensational blackbody spectrum within minutes of first starting to observe, and the results were announced within two months of the launch of *COBE*.[2] The scientific community had to wait another two years to hear the results of the DMR (Differential Microwave Radiometer) experiment to measure the small-scale anisotropy of the microwave background. Ground-based experiments had been attempting to measure these anisotropies for over a decade, and as ever stronger limits were set, theories for the origin of structure in the universe, galaxies and clusters of galaxies, had to be repeatedly modified. As I explained in Chapter 6, from about 1980 onwards it became clear that a universe containing just ordinary baryonic matter, protons and neutrons, and radiation could not be consistent with the very smooth, isotropic background observed. There had to be some other form of dark matter present in the universe that had escaped the control of radiation at a much earlier time and in which density fluctuations could grow, but without imprinting any pattern on the microwave background radiation. By 1990, the consensus was that the dark matter must consist of particles that did not move around at speeds close to the speed of light like neutrinos, and so it had become known as cold dark matter. We all expected that the DMR experiment would detect fluctuations in the microwave background radiation, confirming the cold dark matter scenario, and we were waiting impatiently for the results.

After several false alarms, when announcements were expected at successive cosmology conferences, the *COBE* team was finally scheduled to make its announcement at a NASA press conference on 23 April 1992. This press conference was unfortunately upstaged by an interview given by George Smoot, the leader of the DMR instrument,[3] the previous evening to a local California journalist. This interview found its way to the Associated Press wire service, and cosmologists around the world found their phones ringing from journalists asking what the significance of this announcement was. The front pages of many of the world's leading newspapers, including the *New York Times*, the next morning carried the story that *COBE* had found the primordial seeds from which galaxies and clusters of galaxies have grown. Strictly speaking, the fluctuations measured by *COBE* corresponded to scales today of several hundred million light years, and therefore were on even larger scales than clusters of galaxies. George Smoot's premature press interview caused immense ill feeling within the *COBE* team.

We could now compare the fluctuations measured at 400,000 years after the Big Bang by *COBE*, on very large scales of hundreds of millions of light years, with the structure in the galaxy distribution seen today, on scales of millions of light years, from the *IRAS* galaxy surveys. It was immediately clear that a simple universe with just cold dark matter and ordinary baryonic matter was not going to work. In a sense, we already knew this from our *IRAS* large-scale structure studies.

There were two possibilities to explain the extra structure seen on large scales. One was that neutrinos, particles which pervade the universe and were normally considered to be massless, should have a small nonzero mass. We talked about this idea in Chapter 6 as a possible explanation of dark matter. This would mean two kinds of dark matter: cold dark matter, making up the bulk of the dark matter needed to understand the formation of galaxies and the dark halos of galaxies, and a further ingredient of neutrinos with mass to explain the additional large-scale structure. This became known as the *mixed dark matter* model.

The second possible explanation for the unexplained large-scale structure was that Einstein's cosmological constant should be positive. The cosmological constant was introduced by Einstein in order to produce a static universe in which the self-gravity of the universe is balanced by the repulsive effect of the cosmological constant (see Chapter 6). It can be interpreted as the energy density of the vacuum and has come to be called *dark energy*. Because the cosmological constant plays a part in defining the geometry of the universe, it was possible to test for it through observations of distant objects.

DISTANT SUPERNOVAE FAVOUR THE COSMOLOGICAL CONSTANT

Type Ia supernovae occur when gas is dumped from a companion star onto a white dwarf star in a binary system. The system starts off as two stars similar to the Sun orbiting each other, one more massive than the other. The more massive star goes through its evolution to become a red giant first, then ejects its outer layers to become a white dwarf. Later the second star evolves to become a red giant, but as it does so a wind of gas flows from the red giant onto the white dwarf star. Eventually this pushes the white dwarf above the maximum stable mass it can sustain. The white dwarf star blows up and becomes extraordinarily bright, brighter than its host galaxy for a few months before declining again. These supernovae can be found in very distant galaxies, and because they are similar wherever they occur, they can be used to estimate distance. Two teams embarked on systematic programmes to find supernovae in very distant galaxies using 4-metre ground-based telescopes to detect the supernovae and then following them up with 10-metre telescopes and the *Hubble Space Telescope*. In 1998, both teams announced their results and found that distant supernovae are fainter, and therefore farther away, than expected.[4] The interpretation was that the expansion of the universe really is being accelerated by the repulsive effect of the dark energy.

Curiously, from large underground experiments neutrinos have been found to have a small nonzero mass, so we do in fact live in a mixed dark matter universe. But although we still do not know exactly what those masses are, the contribution of neutrinos to the total matter in the universe appears to be quite small (about 0.1%), too small to affect large-scale structure.

THE INTEGRATED BACKGROUND RADIATION FROM GALAXIES

The last of the three *COBE* instruments, DIRBE, measured the far-infrared and submillimetre spectrum of the Galaxy and also, more significantly, of the integrated background radiation from all galaxies. Across most of the waveband covered by DIRBE, 1.25–240 microns, Mike Hauser and his team were not able to detect the integrated background from galaxies, but at 140 and 240 microns they were able to claim a detection.[5] They concluded that the integrated infrared background energy was about twice that deduced in the optical from deep galaxy counts with the *Hubble Space Telescope* and hence that star formation must be heavily enshrouded in dust at high redshift. The FIRAS team also claimed to have detected the infrared background radiation at 125–2000 microns.[6] The *COBE* teams, however, were a little slow to arrive at these results and had been anticipated by a French group led by Jean-Loup Puget, who had carried out a detailed analysis of the already released FIRAS data and claimed a tentative detection.[7]

THE *WILKINSON MICROWAVE ANISOTROPY PROBE* (WMAP)

The success of *COBE* in detecting the cosmic microwave background fluctuations inspired a new cosmic microwave background mission, which became the *Wilkinson Microwave Anisotropy Probe* (WMAP). Proposed to NASA in 1995, it was selected for the NASA Explorer programme in 1996 and launched amazingly quickly on 30 June 2001 aboard a Delta II rocket. The Europeans also started to plan a cosmic microwave background mission, but this took much longer to bring to fruition and was finally launched as the *Planck* mission in 2009 (see Chapter 12). *WMAP* was despatched to a point a million miles from Earth in the direction away from the Sun, known as the Sun-Earth 2nd Lagrangian (L2) point. This is a stationary point in the gravitational field of the Sun-Earth system, and a spacecraft can be made to orbit around it with just a small amount of fuel needed for occasional orbit corrections to stop the spacecraft from drifting away.

In 1772, the French mathematician and astronomer Joseph Louis Lagrange (1736–1813) analyzed the gravitational field around two bodies orbiting each other, like the Sun and Earth. He found there were five points at which a test mass, such as a satellite, could sit without falling towards either body. These are known as the Lagrangian points. The first one lies between the Sun and Earth and is where the attraction of the Sun and Earth balance each other out. In fact, because the line joining the Sun and Earth is rotating through the year, we also have to take into account centrifugal force in locating this point. The second Lagrangian point lies on the opposite side of the Earth from the Sun, about a million miles from Earth, and here the attraction of the Sun and the Earth are exactly balanced by centrifugal force. It was this point about which *WMAP* was made to orbit. The 3rd Lagrangian point lies on the opposite side of the Sun from the Earth, and the 4th and 5th points are close to the Earth's orbit (Figure 8.2). The advantages of L2 for infrared and submillimetre missions are that it is well away from the Earth's thermal radiation, it is a very stable environment, and the entire sky can be viewed over a six-month period as the Earth orbits the Sun.

WMAP completed its first year's survey in September 2002, and the *WMAP* team released a set of 13 scientific papers in February 2003. *WMAP*'s detailed map of the microwave sky, after subtraction

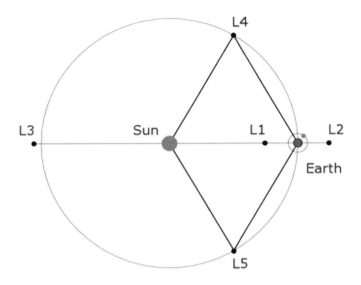

Figure 8.2. Schematic sketch of the Sun-Earth system showing the Lagrangian points. *WMAP* was placed in orbit around the Sun-Earth L2 point.

of foreground radiation from the Galaxy and correction for the dipole anisotropy caused by the Galaxy's motion through the cosmic frame, gave a wonderful insight into the early universe (Plate VII). The detailed pattern of fluctuations on different scales tells us the composition of the universe. The main fluctuation scale is an echo of the Big Bang itself, frozen at the moment that the energy density of radiation fell to be equal to that of matter. Other smaller angular scales seen are harmonics of this, and just as we can identify a musical instrument from the pattern of harmonics it produces, we can estimate the contribution of different types of matter and energy to the total for the universe from the harmonics in the microwave background structure.

The *WMAP* team found the curvature of space to be within 1% of a flat, Euclidean space. They estimated the age of the universe to be 13.7 billion years, again to an accuracy of 1%. And they were able to estimate the composition of the universe to high precision, with about 4% contributed by ordinary atoms (baryons) and 23% made up of some kind of nonbaryonic dark matter.[8] While we do not know what this is comprised of, we do have some ideas about it from modern particle physics theories. Searches for the particles making up the dark matter are going on in underground laboratories and at the CERN Large Hadron Collider. The next few years may see the detection of the dark matter particles that make up most of the matter in the universe. Finally, the remaining 73% of the universe is made up of the mysterious dark energy, which acts to make gravity a repulsive force on the largest cosmological scales. Although many independent experiments point towards the existence of this constituent of the universe, little progress has been made in understanding what it is and why the universe has embarked on a new phase of accelerated expansion, driven by the dark energy. Several ground-based galaxy surveys and new space missions to carry out galaxy surveys are being designed to try to characterize the dark energy and test whether it changes with time.

THE DARK AGES

WMAP also succeeded in measuring the *polarization* of the microwave background radiation, showing that the light waves vibrate preferentially in certain directions perpendicular to our line of sight rather than others. The vibrations of light occur in the plane perpendicular to the direction the light is propagating, and normally these vibrations occur in all possible directions. But when light is reflected from a surface, the vibrations parallel to the surface are reflected more strongly, so the reflected light now vibrates in a preferred direction: the light has become 'polarized'.

The polarization gives information, through detailed modelling, on when the hydrogen in the universe became ionized for the first time after the end of the hot Big Bang phase. The period of the universe from the moment 400,000 years after the Big Bang, when we see the microwave background, to the earliest times at which we see galaxies and quasars, about 600 million years after the Big Bang, is referred to as the 'dark ages'.

In terms of the redshift, which is a useful way of characterizing the evolution of the universe, the 'reionization' of the universe seems to have happened at a redshift between 17 and 11, corresponding to times between 400 and 600 million years after the Big Bang. Theorists are trying to understand whether there is enough ultraviolet radiation from the formation of the first stars and galaxies to have caused this reionization. So far, astronomers have managed to detect quasars and galaxies at redshifts 6–8 (0.8–1.1 billion years after the Big Bang) but have found it harder to push further back in time. A major goal of the next generation of infrared telescopes both in space and on the ground (see Chapter 12) is to detect the first light from stars and galaxies, expected to be emitted at redshifts greater than 10 and perhaps at redshifts as high as 30, only 200 million years after the Big Bang. The evolution of the universe is summarized schematically in Plate VIII.

COBE and *WMAP* have moved us forward into an era of precision cosmology, and we seem to have a consensus amongst cosmologists on the nature of the universe we inhabit, though we still do not understand how it has come to be this way. Ordinary matter makes up only 4% of the universe, cold dark matter contributes 23% and the unexplained dark energy makes up 73%. The time since the Big Bang has been measured as 13.7 billion years, with a precision of 1%. I find it remarkable to have seen cosmology change from a totally speculative field in 1964, before the discovery of the cosmic microwave background, to the post-*WMAP* era of a precision science.

After this detour into the very earliest moments of the universe, we now return to the post-*IRAS* world of infrared astronomy, which would at first be dominated by new, large, ground-based telescopes.

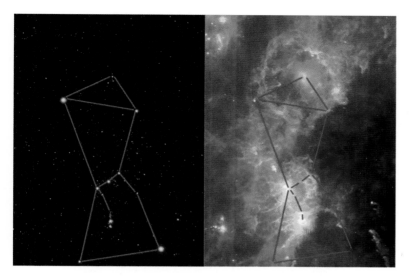

Plate I.   Right: Infrared image of the constellation Orion from the *Infrared Astronomical Satellite* mission (see Chapter 7) compared with an optical image on the left. The lower of the two bright patches in the lower centre of the infrared image is the Orion Nebula (Messier 42) in the Sword of Orion, while the bright nebulosity just above it and to the left surrounds the Belt star Zeta Orionis. The bright spot surrounded by a large ring is Lambda Orionis, and the spot just outside this ring on the left is Betelgeuse.

Plate II.   William Herschel.

Plate III.   Artist's impression of William Herschel conducting his 1800 experiment in which he shows that the Sun radiates invisible radiation beyond the red end of the visible spectrum.

Plate IV.   Artist's impression of a supermassive black hole in the centre of an active galaxy. Twin jets of radio emission emerge from the black hole, which is surrounded by a swirling disk of hot gas. Farther out there is a doughnut-shaped torus of dust, which absorbs light from the hot disk and reradiates it at mid-infrared wavelengths.

Plate V. Point sources from the *IRAS* all-sky survey, colour coded with blue for 12 microns, green for 60 microns and red for 100 microns. The white dots along the plane of the Milky Way are regions of massive star formation, the band of blue dots are red giant stars, the red streaks are clouds of interstellar dust, and the green dots, spread over the whole sky, are galaxies.

Plate VI. *Hubble Space Telescope* image of the rich cluster of galaxies Abell 2218. Many bright elliptical galaxies belonging to the cluster can be seen. Also visible are numerous circular arcs, which are gravitationally lensed images of star-forming galaxies far behind the cluster.

Plate VII. Map of cosmic microwave background fluctuations on the sky made by *WMAP*. The dipole anisotropy and the emission from the Milky Way have been removed. Hotter than average directions are shown in yellow or red, colder directions in blue. The contrast has been enormously exaggerated: these fluctuations are only of order 0.001%.

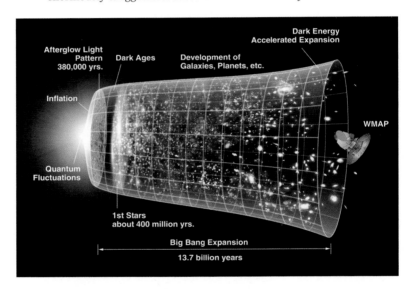

Plate VIII. The history of the Big Bang. A two-dimensional slice of the universe is shown evolving with time from left to right. The universe starts off as some kind of quantum fluctuation and goes through a brief phase of exponential inflation. At 400,000 years after the Big Bang, the universe becomes transparent, a phase we see now as the microwave background radiation. The 'dark ages' follow, and the first stars form about 400 million years later. Galaxies start to accumulate and with time aggregate into clusters. In the final few billion years, the expansion of the universe starts to accelerate under the influence of dark energy.

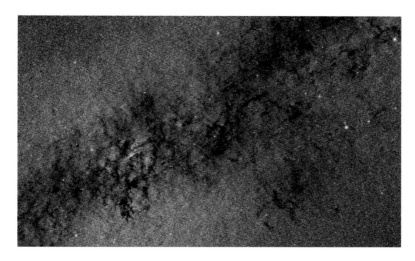

Plate IX.   2MASS infrared image of the Galactic centre at 1.25–2.2 microns. Although the star clusters near the centre of the Galaxy begin to be seen, the obscuration by dust remains very strong.

Plate X.   The James Clerk Maxwell submillimetre telescope on Mauna Kea, Hawaii.

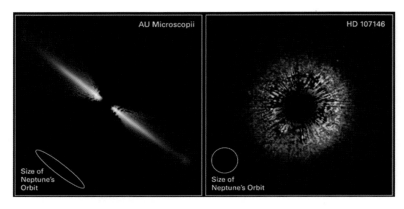

Plate XI.   Debris disks around two stars, one edge-on and the other face-on, imaged by the *Hubble Space Telescope*.

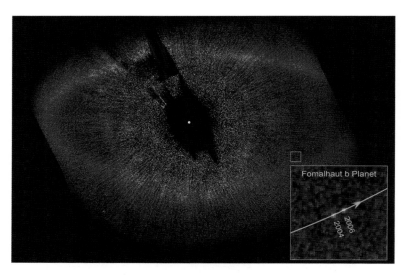

Plate XII.   *Hubble Space Telescope* image of the debris disk around the bright star Fomalhaut, which was first detected by *IRAS*. Inset shows the location of an exoplanet detected just inside the main dust ring.

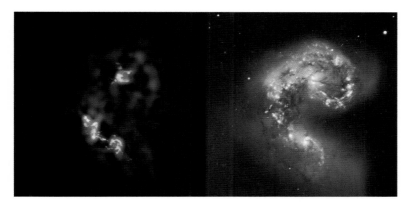

Plate XIII.   *Hubble Space Telescope* image of the 'Antennae' galaxies, two strongly interacting galaxies in the process of merging together (right panel). The ALMA millimetre image (left panel) shows that most of the energy in the starburst is hidden from view in the optical part of the spectrum.

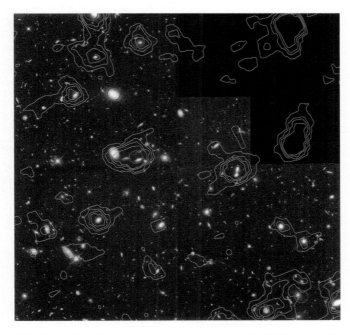

Plate XIV.   The Hubble Deep Field, imaged by the *Hubble Space Telescope* in 1996, with contours of 15-micron brightness overlaid from the *ISO* survey (Aussel et al. 1999).

Plate XV. A composite image of nearby galaxies imaged with the *Spitzer* space telescope by the SINGS collaboration, arranged in the form of Hubble's 'tuning fork' diagram, with elliptical galaxies forming the handle of the tuning fork and normal and barred spiral galaxies forming the upper and lower prongs. Irregular galaxies have been placed to the lower left.

Plate XVI. *Spitzer* infrared image of Messier 51, the 'Whirlpool' Galaxy. The colour coding in these *Spitzer* images has blue representing the shortest wavelength, 3.6 microns, and red representing 24 microns.

Plate XVII. *Spitzer* image of Messier 81.

Plate XVIII. The nearby starburst galaxy Messier 82 as seen in the optical (upper panel) and infrared (lower panel).

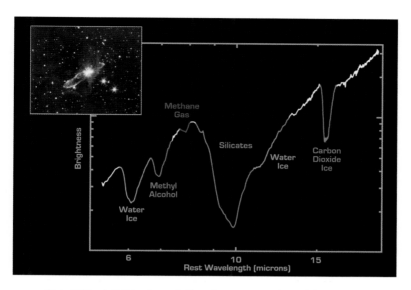

Plate XIX.   A 5–20-micron infrared spectrum taken with the *Spitzer* space telescope of the young stellar objects Herbig-Haro 46 and 47 (image inset). Absorption features caused by water and carbon dioxide ice, methyl alcohol, methane gas and silicate dust grains can be seen.

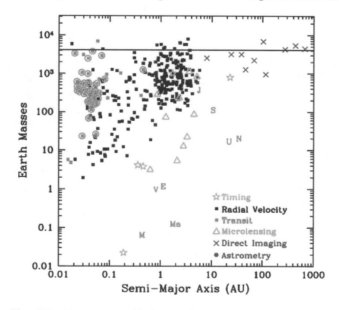

Plate XX.   Mass versus orbital radius for 300 known exoplanets. The locations of Solar System planets are indicated in red lettering. The colour coding for exoplanets corresponds to different methods of detection. Points circled in red are those which have been studied with *Spitzer*.

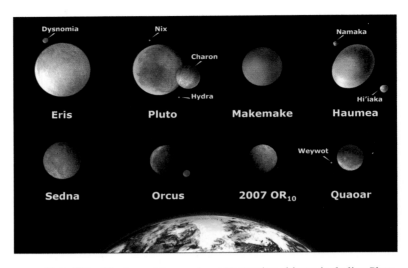

Plate XXI. The largest known trans-Neptunian objects, including Pluto (plus the largest asteroid, Eris).

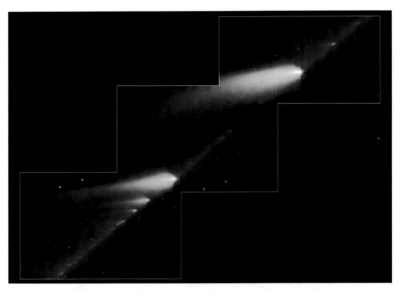

Plate XXII. Debris from Comet Schwassmann-Wachmann 3 spread along the orbit of the comet, blown away from the Sun by the action of radiation pressure and the solar wind, imaged here in infrared light by *Spitzer*.

Plate XXIII.   Artist's impression of the *Herschel* satellite. Below the 3.5-metre mirror is the cryostat containing the instrument package.

Plate XXIV.   Launch of *Herschel* and *Planck* from Kourou, French Guiana, on 14 May 2009.

Plate XXV. An early *Herschel* submillimetre image of part of the Milky Way, in the constellation of Vulpecula, made by combining SPIRE and PACS data.

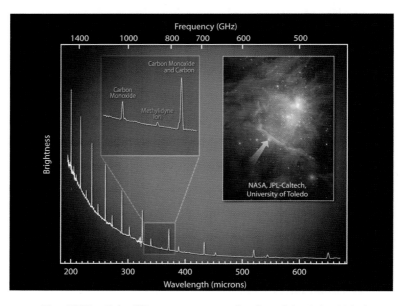

Plate XXVI. Submillimetre spectrum of region of the Orion Nebula (image inset) taken with the SPIRE spectrometer, showing a wealth of molecular lines.

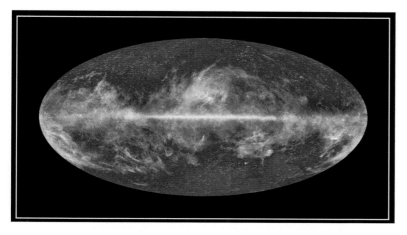

Plate XXVII. *Planck* submillimetre map of the whole sky after the first year of data. The blue and white show the emission from the Galaxy, the red and orange the fluctuations in the cosmic microwave background.

Plate XXVIII. *Planck* image of a section of the Milky Way. We're mainly seeing dust within 500 light years of the Sun, but the white strip is hotter, more distant dust heated by newly forming stars.

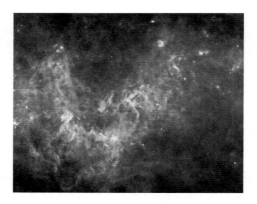

Plate XXIX.   *Herschel* image of a portion of the Milky Way.

Plate XXX.   *Herschel* image of a dark cloud in the Southern Cross.

Plate XXXI.   *Herschel* image of a portion of the Rosette Nebula. The bright stars illuminating this dust cloud are off to the right.

9

Giant Ground-Based Near-Infrared and Submillimetre Telescopes

The enormous impact of the Caltech Two Micron Survey in 1969 stimulated interest in developing specialized ground-based infrared telescopes. While far-infrared wavelengths are accessible only from space, atmospheric windows can be used in the near-infrared and submillimetre wavelengths from high mountaintop sites. From the late 1970s, large near-infrared and submillimetre telescopes began to be built on high-altitude sites, especially on the 4200-metre dormant volcanic peak of Mauna Kea, Hawaii.

The advent of these new infrared telescopes and the dramatic impact of the new infrared-array detectors in the 1980s generated a profusion of scientific discoveries. These ranged from high-redshift galaxies to studies of luminous dusty galaxies and the effects of gravitational lensing, the black hole in the heart of our Milky Way Galaxy, brown dwarf stars, protostars and protoplanetary systems.

The first of the large, specialized infrared telescopes was the United Kingdom's 3.8-metre UK Infrared Telescope (UKIRT), constructed between 1975 and 1978 on Mauna Kea, Hawaii. UKIRT was designed to be lightweight and cheap. It started work in 1979 and had an immediate impact on near-infrared astronomy. At about the same time, NASA's 3-metre Infrared Telescope Facility (IRTF), with Eric Becklin as its first director, and the Canada-France-Hawaii 3.6-metre optical/infrared telescope also began work on Mauna Kea. With their excellent high-altitude sites and careful design of their optics to keep stray heat radiation from the support structures from being seen by their detectors, UKIRT and IRTF were for over a decade the most powerful telescopes at near-infrared wavelengths.

The single technical development that did most to put these infrared telescopes on the map was the development of infrared-array detectors. The *IRAS* mission had used an array of 62 detectors, but in the late 1970s much larger infrared arrays, typically with 32×32 detectors, began to be built by several different companies, especially Santa Barbara Research Center and Rockwell International, mainly for military applications. Some of these began to be used by U.S. astronomers, notably Judith Pipher and Bill Forrest at the University of Rochester. Forrest developed an array for IRTF and with Giovanni Fazio and others started to develop an infrared-array camera (IRAC) for what was to become the *Spitzer Space Telescope* (see Chapter 10). Don Hall, Eric Becklin and others used a 64×64 infrared array developed by Rockwell and the Jet Propulsion Lab at IRTF. French astronomers also began to use a 32×32 infrared array and went on to develop an infrared array (ISOCAM) for the *Infrared Space Observatory* (Chapter 10). The first observatory infrared array available for any observer at the telescope was the IRCAM 58×62 indium antimonide array designed for UKIRT in 1984 by a team at the Royal Observatory Edinburgh (ROE) led by Ian McLean[1] and built by the Santa Barbara Research Center in California. Suddenly, stunning near-infrared images of the sky could be generated in a few minutes of observation.

At a 2009 ROE Workshop, Ian McLean wrote:

> Many results from IRCAM were presented at the March 1987 conference in Hilo called 'Infrared Astronomy with Arrays'.... There is no doubt from the reviews at that meeting that the quality of the IRCAM images caught everyone by surprise. I am often asked whether or not I knew that the world of infrared astronomy had changed. I think we did have a good sense that something major had just occurred and that UKIRT was right at the center of it. Based on comments from Wayne van Citters (National Science Foundation) and Don Hall (University of Hawaii) at the close of the Hilo meeting, I think the community also realized that a turning point had been reached. Wayne van Citters said: 'I don't think there's any doubt that we are on the verge of a promised land in optical and infrared astronomy.' Referring specifically to IRCAM, Don Hall summarized the conference by saying, 'Things that seemed in the future and beyond my grasp for so long are clearly here.'

> The buzz and excitement of the conference bubbled over and was captured in a moment that will live with me forever. In the middle of my talk I showed an IR-CAM image of OMC-2 obtained by John Rayner and me. As I displayed the polarization map of the reflection

nebula and then a set of images at progressively longer wavelengths to illustrate the changes in appearance with less reddening I said, '*If this is what infrared astronomy is going to be like from now on, then all I can say is – I like it!*' To my astonishment I received spontaneous applause in the middle of the talk. Clearly, I struck a nerve, and I think I simply stated what everyone was feeling. It was a great moment.

THE ADVENT OF THE GIANT GROUND-BASED TELESCOPES

In 1985, the Keck Foundation donated $70 million to Caltech for the construction of a 10-metre telescope. This would be twice the diameter of the largest existing fully functioning optical telescope, the Mount Palomar 5-metre (200-inch) telescope, built in 1948. The Keck Telescope, developed jointly by Caltech and the University of California, was completed in 1993 on Mauna Kea, Hawaii. It had a novel segmented mirror design and was the precursor of a dozen or so 8–10-metre class telescopes built over the next decade.[2] These operate at both optical and near-infrared wavelengths, with typically half of their instrument suites being devoted to near-infrared imaging and spectroscopy. The Keck Observatory has two 10.2-metre telescopes on Mauna Kea, the second completed in 1996. The European Southern Observatory inaugurated four 8.2-metre telescopes at Paranal, Chile, between 1998 and 2001, called the Very Large Telescope, or VLT, which can operate independently or in tandem. The Japanese Subaru 8.2-metre telescope was completed on Mauna Kea in 1999 and named after the Japanese word for the Pleiades, and in the same year the Gemini Observatory's 8.1-metre telescope started work also on Mauna Kea, with a second at La Serena, Chile, in 2001. The combination of the 4–5 times larger collecting area compared with the previous generation of 4-metre telescopes and the high altitude of their sites represented a huge advance in sensitivity at optical and near-infrared wavelengths.

An important technical advance was the development of large-format, million-pixel, infrared-array detectors during the 1990s, which emerged just in time for the new 8–10-metre telescopes. The success in the 1980s of the smaller arrays discussed above in astronomical applications led to an effort funded by the National Science Foundation and NASA to develop much larger arrays with over a million pixels.[3] While optical arrays use charge-coupled device (CCD) technology, similar to that found in digital cameras, infrared arrays consist of a dual layer, with the tightly packed grid of infrared sensors forming one layer and a second backing layer of silicon providing the electronic readout.

These large arrays have transformed infrared astronomy, both for large ground-based telescopes and for space missions like the *Hubble* and *Spitzer* space telescopes. Ian McLean, who had led the development of IRCAM on UKIRT, and Eric Becklin moved to the University of California at Los Angeles in 1989 to create the Infrared Lab there and develop infrared instruments for use on the Keck Telescope.[4]

The performance of large ground-based telescopes has been enormously enhanced by the development of adaptive optics to monitor the fluctuations in the Earth's atmosphere and correct for them. These giant ground-based telescopes always have to compete with the *Hubble Space Telescope* (*HST*). Although the *HST* is only a 2.4-metre telescope, its location in space makes it competitive for imaging programmes with the giant ground-based telescopes. In 1997, the astronauts carrying out the second servicing mission to the *Hubble Space Telescope* installed the NICMOS near-infrared 256×256 array camera, which had been developed by Roger Thompson and his colleagues at the University of Arizona, on the *HST*.[5] A new wide-field ultraviolet, optical and near-infrared camera, WFC3, was installed on the *HST* in May 2009 and is producing some spectacular images.

2 MASS AND OTHER INFRARED SURVEYS

In 1997, the University of Massachusetts and California Institute of Technology embarked on a project to make a new near-infrared survey of the whole sky, the first since the 1969 Two Micron Survey, known as the Two Micron All-Sky Survey (2MASS). The entire sky was surveyed in the J (1.25 microns), H (1.65 microns) and K (2.2 microns) bands using two highly automated 1.3-metre telescopes, one at Mount Hopkins, Arizona, and the other at Cerro Tololo (CTIO), Chile. The final all-sky data comprised a digital atlas of the sky, a point-source catalogue consisting of 300 million stars and other unresolved objects, and an extended-source catalogue with over one million galaxies and other nebulae, and was released in 2003 (see Figure 9.1 and Plate IX).[6]

Other large near-infrared surveys include DENIS, a deep survey carried out by French astronomers in the I (0.8 micron), J and K bands of the southern sky between 1996 and 2001,[7] the CFHT Legacy Survey, carried out at the Canada-France-Hawaii Telescope between 2003 and 2009,[8] and the UKIRT Deep Sky Survey (UKIDSS), which began in 2003 and is still continuing.[9]

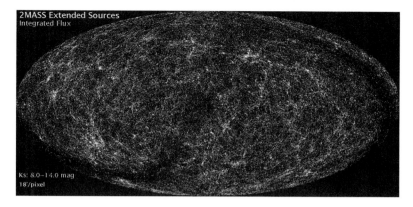

2MASS Extended Sources
Integrated Flux

Ks: 8.0–14.0 mag
18'/pixel

Figure 9.1. The 2MASS all-sky galaxy survey. Each dot represents a galaxy, and many clusters of galaxies and filamentary structures can be seen.

HIGH-REDSHIFT GALAXIES

One of the focuses of observational cosmology has always been to try to reach out to the most distant objects in the universe. In the past decade, this focus of extragalactic research has shifted towards galaxies with redshifts 5–8. Their starlight is shifted entirely to infrared wavelengths, hence the focus on new deep infrared surveys. Many of the high-redshift galaxies found to date have been detected using the *Hubble Space Telescope*, which in addition to its amazing imaging performance at optical wavelengths has the near-infrared camera NICMOS. Searches involve looking for faint objects detected in the near-infrared bands that are blank at optical wavelengths even in the deepest integrations.[10] The highest-redshift galaxy detected to date, with redshift 8.6 (800 million years after the Big Bang), was found in deep infrared images with the *Hubble Space Telescope*, and its redshift was measured using the SINFONI infrared spectrograph on the VLT.[11] The highest-redshift quasar found to date, with redshift 7.1, was found in the UKIDSS survey.[12]

THE STAR-FORMATION HISTORY OF GALAXIES

The advent of large-format arrays at optical and near-infrared wavelengths opened up the possibility of large galaxy surveys, and these have been used to study the star-formation history of the universe. Simon Lilly led a large survey of several thousand galaxies, initially selected in the I band at 0.83 microns and then followed up at visible

(0.44 and 0.55 microns) and infrared (2.2 microns) wavelengths.[13] Lilly and his colleagues used this survey to estimate the star-formation history of the universe between redshifts 0 and 1 and found that the rate of star formation in galaxies increases steeply as we look back into the past.[14] Piero Madau and his collaborators used a deep survey with the *Hubble Space Telescope*, in a region called the Hubble Deep Field, to extend this star-formation history to redshift 4, finding that the star-formation rate peaked at around redshift 2 and then declined towards higher redshifts.[15]

In contrast to these and other[16] surveys of very distant galaxies, 2MASS has been used to map out our nearest neighbours, dwarf galaxies orbiting our own Galaxy, which are very hard to see against the foreground stars of the Milky Way.[17]

THE SUPERMASSIVE BLACK HOLE AT THE CENTRE OF OUR GALAXY

The central regions of the Galaxy have long been a subject of intense interest. The group at the University of California at Berkeley, led by Nobel laureate Charles Townes, studied ionized gas in the centre of the Galaxy throughout the late 1970s and 1980s, drawing attention to the high velocities seen in this gas, which indicated a concentrated mass of several million solar masses in the Galactic centre.[18] Could this be a black hole, a miniature version of the huge monsters which power quasars? In their 1987 review, Reinhard Genzel and Townes concluded that the evidence for a central million solar mass black hole was 'substantial but not fully convincing'.[19] Genzel then moved back to Germany and built up a powerful infrared group at the Max Planck Institute for Extraterrestial Research in Munich, specializing in state-of-the-art infrared instrumentation. In 1997, Genzel reported the first results of a four-year programme to measure the motions of individual stars across the sky ('proper motions') in the innermost core of the Galactic centre using high-resolution near-infrared imaging at the ESO 3.6-metre New Technology Telescope (NTT) in La Silla, Chile.[20] Monitoring the positions of the stars over several years gives their orbits on the sky, and the mass of the central object they are orbiting can be deduced. The central mass was now estimated as 2.5 million solar masses, and the very small size it would require pointed strongly towards it being a black hole. Genzel's group continued to monitor the proper motions of the stars near the centre of the Galaxy and were able to trace them around substantial portions of their orbits, demonstrating rather conclusively that the Galaxy contains a supermassive black hole at its centre.[21] A group at the University of California at Los Angeles, led by Andrea Ghez, pursued a similar proper-motion programme using the Keck Telescope and in 1998 announced very similar results.[22] While the black holes in some active galaxies are one hundred or a thousand times more massive than the one at the centre of our Galaxy, the proximity of our own nuclear black hole allows us to follow the motions of stars orbiting around it and demonstrate that there really must be a black hole there.

BROWN DWARF STARS

Brown dwarfs were first detected in the 1990s and have been the subject of detailed study ever since. They are objects with masses between 10 and 80 times that of Jupiter (i.e., between 1% and 8% of the mass of the Sun) and therefore massive enough to have an episode of nuclear burning of their primordial deuterium, left over from the Big Bang, but not massive enough to reach the point of hydrogen burning (Figure 9.2). They are too big to be planets and can be thought of as failed stars. In the neighbourhood of the Sun there are probably as many brown dwarfs as there are all other types of stars.

Brown dwarfs have been classified, on the basis of their spectra, into classes L, T and Y. L-class objects have temperatures in the range 1400–2600 K, have masses below the hydrogen-burning limit (though they do have a brief phase of deuterium burning) and have no methane in their atmospheres. T-class objects have temperatures in the range 500–1400 K and have methane in their atmospheres, which strongly absorbs their infrared spectra. The Y-class objects are expected to have temperatures in the range 150–500 K and to be cool enough to have water clouds. The letters L, T and Y were carefully selected to add to the standard OBAFGKM sequence of stellar spectral types, as most other letters of the alphabet were either already in use for types of stars or could have a confusing meaning.[23]

The 2MASS survey had a major impact on studies of brown dwarfs. Prior to 2MASS, just six brown dwarfs were known, all of class L. The first 400 square degrees surveyed by 2MASS in 1999 yielded 20 new brown dwarfs, including the first of class T.[24] Within a year, a further 67 L-class brown dwarfs had been found by 2MASS. The newly discovered brown dwarfs were followed up by taking infrared spectra with the Keck Telescope. In 2007, Steve Warren and his collaborators found the nearest cool brown dwarf known at that time, with a temperature

Figure 9.2. Relative sizes of a low-mass star, a brown dwarf, Jupiter and the Earth.

of about 600 K, using the UKIDSS survey,[25] and even cooler brown dwarfs have been found with the *WISE* mission (see Chapter 12). Over one thousand brown dwarfs have been found to date.

During the 1980s there was speculation that brown dwarfs could make up a major fraction of the mass of the Galaxy and could even account for the dark matter whose presence was inferred from the orbital speeds of stars around the Galaxy. But although some of the Sun's nearest neighbours may turn out to be brown dwarfs it is not now considered likely that they make up a significant fraction of the matter in the Galaxy.

So far we have talked about the discoveries made with the new ground-based telescopes working in the near-infrared windows at wavelengths of 1–3 microns. We now turn to the long-wavelength end of our infrared to submillimetre band, at wavelengths of 300 microns to one millimetre.

SUBMILLIMETRE TELESCOPES

The early work in millimetre astronomy was carried out in the 1970s with the NRAO 11-metre telescope on Kitt Peak, Arizona (Chapter 5).

This was converted to a 12-metre telescope in 1987 and continued to work until 2000. A millimetre-wave interferometer began to be developed in the late 1970s by Jack Welch at Berkeley's Hat Creek Observatory (BIMA). An interferometer works by combining the signal from two or more separate antennas and using the difference between the signals to mimic the resolving power of a telescope as big as the space between the antennas. Caltech also developed a millimetre-wave interferometer at Owens Valley Radio Observatory (OVRO) in the late 1970s. The University of Massachusetts and four other colleges built a 14-metre millimetre-wave telescope in 1976, and Swedish astronomers built the 20-metre millimetre-wave telescope at Onsala at about the same time. The European Instituto de Radioastronomia Millemetrica (IRAM) was founded in 1979 and operates the 30-metre millimetre telescope in the Spanish Sierra Nevada at an altitude of 3000 metres. The observatory is run jointly by France, Germany and Spain, with its headquarters in Grenoble. It operates at wavelengths of 1.0, 1.3, 2 and 3 millimetres and can perform both molecular-line observations and continuum observations (in which the radiation is spread broadly over a wide range of wavelengths). There is also an associated millimetre interferometer at the Plateau de Bure. The Japanese Nobeyama Radio Observatory operates a 45-metre millimetre-wave telescope, completed in 1982, and the Nobeyama Millimetre Array.

These observatories all operate mainly outside the submillimetre band, at wavelengths 1 millimetre or longer. The first dedicated submillimetre observatories were the James Clerk Maxwell Telescope (JCMT, Plate X), built by the United Kingdom and the Netherlands, and the Caltech Submillimeter Observatory (CSO). The two telescopes sit next to each other in 'submillimeter valley' near the summit of Mauna Kea in Hawaii, and both started operations at around the same time, in 1986. Canada joined the JCMT in 1988. The Swedes, in collaboration with the European Southern Observatory, built the SEST 15-metre submillimetre telescope at La Silla, Chile, in 1987.

The Caltech Submillimeter Observatory (CSO) is a 10.4-metre submillimetre wavelength telescope that began operations in 1986 and will be decommissioned in 2016 as the new CCAT (Cerro Chajnantor Atacama Telescope) comes into operation in Chile. The CSO was designed and built by Bob Leighton and was operated under the leadership of Tom Phillips.[26] I have given some of the background of the United Kingdom's James Clerk Maxwell Telescope (JCMT) in Chapter 5. Tom Phillips was responsible for setting the JCMT project on its way, but the key person in bringing the telescope to fruition was Richard

Hills of Cambridge University, who was the project scientist for the telescope. The JCMT is a 15-metre telescope that can operate from 300 microns to 2 millimetres, with both continuum and molecular-line receivers, and began operation in 1986. For the first time it was possible to study the submillimetre spectra of Galactic and extragalactic sources, and even to build up a point-by-point map, but with the original JCMT submillimetre photometer this was slow work.

SUBMILLIMETRE GALAXIES

The SCUBA bolometer array,[27] completed in 1996 and installed on the JCMT, was the first proper submillimetre camera and allowed rapid mapping of Galactic and extragalactic sources. Prior to that, small regions had to be mapped painstakingly using single-detector instruments. SCUBA had 128 detectors, operating at a temperature of 0.1 K for extra sensitivity, and provided a huge advance in mapping speed.

With colleagues from Edinburgh and Cambridge, I proposed a survey of an area of the sky that had been mapped deeply with the *Hubble Space Telescope*, the Hubble Deep Field. We were lucky enough to get probably the two best weeks of weather in the history of the JCMT for the initial observations. It was exciting to be at the telescope for one of the early observing runs with SCUBA, with a sensitivity one thousand times better than the detector system my Queen Mary College colleagues and I had used in the 1970s.

Working at 14,000 feet has its own hazards. The air is thin and it is best to do everything very slowly. The brain does not seem to function properly. The last section of the road to the summit of Mauna Kea is unpaved and precipitous. I always feel completely terrified driving along it, and there have been a few deaths from cars leaving the road. At the summit, the volcanic rock is red and dusty, like the surface of Mars. There are no plants or animals. The telescopes, however, are beautiful and the night sky unbelievable.

Amazingly there were sources in our submillimetre map of the Hubble Deep Field (Figure 9.3), and it was soon clear that these had to be high-redshift and very luminous galaxies. The announcement of our results in *Nature*, and of the new class of submillimetre galaxies, had a tremendous impact.[28] From *IRAS* observations of galaxies and follow-up with the *Kuiper Airborne Observatory* at longer wavelengths, we knew that the peak wavelength for galaxy far-infrared spectra lay in the range of 50–200 microns. For distant galaxies with redshifts greater than 3, the redshift brought this peak into the 850-micron

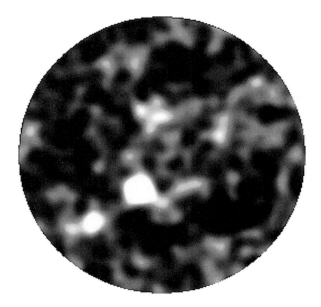

Figure 9.3. The 850-micron survey of the Hubble Deep Field carried out
with the SCUBA bolometer array on the JCMT.

band, so such galaxies could appear very bright at submillimetre
wavelengths. This had been pointed out by Alberto Franceschini of
the University of Padova and his co-workers in 1991.[29] Subsequently
we carried out a larger submillimetre survey,[30] and Jim Dunlop then
led a large consortium in the even larger 'SHADES' survey.[31] Consortia
led by Steve Eales have carried out submillimetre surveys both of random areas of sky and of known nearby infrared galaxies.[32] The rate of
progress was still quite slow, with about one source detected per night
of observation. With the new JCMT submillimetre bolometer array,
SCUBA2, with over 10,000 detectors, submillimetre surveys will finally
become easy.

One technique that made searches for submillimetre galaxies a
bit easier was to point the telescope at rich clusters of galaxies. The
galaxies in the cluster act as a gravitational lens, amplifying the light
from distant galaxies behind the cluster (see Plate VI).[33] Several submillimetre galaxies have been found in these lensing surveys. The
magnification afforded by the gravitational lens allows intrinsically
fainter galaxies to be detected. A difficulty in identifying submillimetre sources is the rather inaccurate positions furnished by submillimetre telescopes, and one trick to improve this is to look for radio sources

at the same location. The more accurate positions of the radio sources can then be used to identify the correct optical galaxy for study with 8–10-metre optical telescopes, with which we can determine the redshift. In this way, the redshift distribution of submillimetre galaxies has been estimated.[34] Most are in the range 1–4, with a peak around 2.5. There has been much discussion of the nature of submillimetre galaxies, emphasizing both their unique nature, perhaps as the precursors of giant elliptical galaxies, and their continuity with other types of infrared galaxies.[35] One interesting question has been whether they are mainly associated with quasars, and hence perhaps powered by black holes rather than by star formation, but no very convincing link has been found. There is no very strong correlation with x-ray emission, for example.[36] Molecular gas, especially carbon monoxide, in high-redshift star-forming galaxies has been studied with IRAM and Nobeyama.[37] Dust in distant submillimeter galaxies and quasars has been imaged with the Plateau de Bure millimetre interferometer[38] and with the SHARC-2 350-micron photometer at CSO.[39] All these studies point towards submillimetre galaxies being luminous starbursts.

The submillimetre galaxies, at high redshifts and with very high luminosities, posed a problem for theories of galaxy formation. Groups at Durham and Munich had made great progress in computer simulations of the aggregation of dark matter fluctuations in the early universe under the action of gravity, the subsequent infall of ordinary (baryonic) matter into these dark matter concentrations, and the bursts of star formation triggered by the mergers of these primeval galaxies. These calculations take account of the controlling effect on star formation that supernova explosions from massive stars, and radiation from supermassive black holes in galactic nuclei, may have. However, the profusion of very dramatic starbursts at high redshifts was unexpected, and the theoreticians have had some difficulty incorporating them into their syntheses.[40] It appeared that such bursts would form too many low-mass stars. One solution is to suppose that these bursts made primarily short-lived high-mass stars.

DUST DEBRIS DISKS

Following the discovery of debris disks around Vega and other stars by *IRAS* in 1984, there were many attempts to image these disks directly. An early success was Beta Pictoris, an edge-on disk that was imaged with the *Hubble Space Telescope* by blocking out the light from the central star (see Plate XI). Over 20 debris disks have now been imaged, many through scattered light at optical wavelengths. However, an important breakthrough was made with the JCMT in 1998 when the debris disks around Vega, Beta Pictoris, Fomalhaut and Epsilon Eridani were imaged at 850 microns.[41] These submillimetre maps demonstrated that the dust emission was coming from regions the size of the Solar System's Kuiper Belt (see Chapter 11). In the case of Fomalhaut, they demonstrated that there was a central cavity free of dust, which suggested planets had already formed and were clearing the dust in the zone of their orbits. This was later wonderfully confirmed by direct imaging of a Fomalhaut planet with the *Hubble Space Telescope* (see Plate XII).

The large ground-based infrared and submillimetre telescopes have generated an extraordinary wealth of discoveries about brown dwarfs, debris disks, high-redshift galaxies and the black hole at the centre of the Galaxy. They have provided a great route for following up the discoveries of *IRAS*. But ground-based astronomy is limited to the wavelength windows through which light from the universe reaches the ground. The success of *IRAS* stimulated European, American and Japanese astronomers to plan new infrared space observatories that would open a new, rich vein of discovery.

10

The *Infrared Space Observatory* and the *Spitzer Space Telescope*: The Star-Formation History of the Universe

The enormous success of *IRAS* stimulated both the European Space Agency (ESA) and NASA to develop new space infrared observatories that would follow up the wealth of discoveries about the infrared universe made with *IRAS*.

Early in February 1983, the European Space Agency met to select a new medium-sized astronomy space mission. Peter Clegg was able to place on the table at the meeting the first scan around the sky from *IRAS*, and its quality was sufficient to convince the European Space Agency to select the *Infrared Space Observatory* (*ISO*). The idea for a European infrared space observatory had been first proposed in 1979. *ISO* was finally launched in November 1995 with a planned life of 18 months (Figure 10.1). In fact, its helium coolant lasted until April 1998, almost a year longer than expected.

ISO had a camera, ISOCAM, led by Catherine Cesarsky[1] and a spectrometer, SWS, led by Thijs de Graauw,[2] working at the near- and mid-infrared wavelengths (3–20 microns); and a camera, ISOPHOT, led by Dietrich Lemke[3] and a spectrometer, LWS, led by Peter Clegg,[4] working at far-infrared (40–160 micron) wavelengths. The two cameras also had smaller low-resolution spectrometers as part of their capability. The spectrometers of *ISO* were especially powerful in unravelling the nature of the dust around stars and in interstellar space, and in probing young stars in the process of formation. The instrument package was extraordinarily complex, and there were criticisms during *ISO*'s development that too much unproven technology was involved and that there were too many

I have made extensive use of the reviews by Genzel and Cesarsky (2000), van Dishoeck (2004), Werner et al. (2006), Soifer, Helou and Werner (2008) and Wyatt (2008), and of the historical review by Werner (2006). A very detailed account of the history of the Spitzer mission is given in George Rieke's book *The Last of the Great Observatories: Spitzer and the Era of Faster, Better, Cheaper at NASA* (2006).

moving parts that might jam in orbit. It is a tribute to the teams involved that these complex and versatile instruments worked so well in space.

It was planned that the cryostat design would be similar to that of *IRAS*, but the engineers had great difficulty getting the crucial valve that vented liquid helium out onto the mirror and instrument package to work. This was the valve that had failed during testing of *IRAS* but, fortunately, worked in orbit. For *ISO* the valve had to be completely redesigned from scratch. Most of the *ISO* systems worked better in orbit than the original planned specification. For example, the accuracy of the pointing ended up being ten times better than specified. Compared with *IRAS*, the *ISO* sensitivity was 40 times more sensitive at a wavelength of 12 microns, and the spatial resolution, the ability to see fine detail, was 20 times better. The *ISO* orbit was highly elongated, from 1000 kilometres at its nearest point to 70,500 kilometres at its farthest point from Earth. It orbited the Earth once every 24 hours and spent 7 hours each orbit inside the Earth's radiation belts, during which time its instruments were switched off.

GALACTIC SOURCES WITH *ISO*

When stars start to form deep inside molecular clouds, the surrounding gas and dust start to fall towards the dense core, and this infalling gas ends up in a disk around the central core. This disk of gas in turn feeds the central protostar from its inner edge. At this stage the central protostar is hidden from view behind a thick veil of dust extinction, and the only observable tracers of this stage of evolution are the infalling gas and dust. The rotating disk is the basic material from which planets are formed. Infrared spectroscopy is then the key to understanding the different stages of the formation of stars and planetary systems (Figure 10.2). *ISO* gave us the first opportunity to see complete infrared spectra from 2.4 to 200 microns.[5]

The spectra of the youngest and coldest protostellar objects, those for which collapse started less than 10,000 years ago, peak at around 100 microns and are best studied with space-based telescopes like *ISO*. Once outflows start to develop from the newly forming stars and to drive the dense envelopes away, which occurs at around 100,000 years, the objects become detectable in the mid-infrared. The very wide range of physical conditions during these different phases of star formation, with densities from ten thousand to ten million million atoms per cubic centimetre and temperatures from ten to ten thousand Kelvin, results in complex chemistry and a diversity of atomic and molecular excitations.

Figure 10.1. Artists's impression of the *Infrared Space Observatory* (*ISO*).

While ground-based submillimetre observations, especially of carbon monoxide, probe gas at temperatures less than 100 K, and optical atomic lines and near-infrared molecular hydrogen observations probe much hotter temperatures (> 2000 K), *ISO* provided a unique insight into conditions at 200–2000 K.[6] *ISO* was able to quantify the heating efficiency

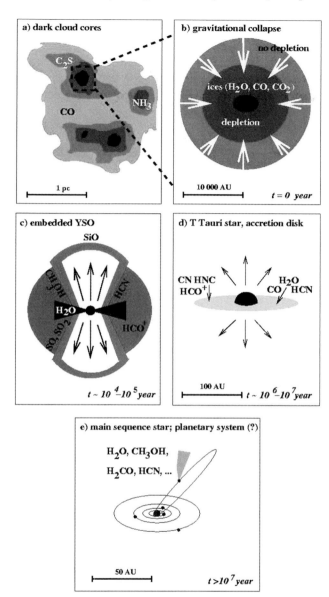

Figure 10.2. Young stellar objects and the locations of the different atoms and molecules whose spectral features were seen by *ISO* (van Dishoeck 2004).

and total cooling power of all the main infrared spectral lines. It also provided an inventory of the main elements present in the gaseous and solid-state phases, especially carbon, nitrogen and oxygen.

ISO reported the first detection of interstellar carbon dioxide, impossible to detect from the ground because of emission and absorption in the Earth's atmosphere. Similarly, *ISO* was able to observe water in a variety of environments, for example the atmosphere of Saturn's moon Titan, again a molecule very hard to observe from the ground because of emission and absorption in our atmosphere. *ISO* also detected absorption by a variety of ices – water, carbon monoxide, carbon dioxide and methane – towards low-mass young stellar objects. Prior to *ISO* the silicate feature in circumstellar dust shells was generally believed to be entirely the result of amorphous silicates, so an exciting discovery of *ISO* was the detection of crystalline silicates in a variety of environments, including dust shells around hot stars, protoplanetary disks and comets.[7] An important mechanism for converting amorphous silicates to crystalline form may be the formation of Jupiter-mass planets within protoplanetary disks.

EXTRAGALACTIC SOURCES WITH *ISO*

IRAS detected huge numbers of galaxies, more than 60,000 in the *IRAS* Faint Source Survey, but most of them were detected only at 60 microns. Only about 1000 galaxies were detected in all four *IRAS* bands at 12, 25, 60 and 100 microns. The large number of photometric bands available with *ISO*, and the sensitive low-resolution spectrometers, permitted the infrared spectra of galaxies to be examined in detail. The dominant effect of the PAH (polycyclic aromatic hydrocarbon) features at 3–13 microns for star-forming galaxies quickly became apparent. The strength of these features was found to be a good tracer of normal and moderately active star formation, whether measured by individual features using the spectrometers or for larger samples of fainter galaxies using the ISOCAM broadband channels centred at 6.7 and 15 microns.[8] At very high radiation intensities, for example in the centre of the Galaxy, in ultraluminous infrared galaxies, or in active galactic nuclei, the strength of the PAH features plummets. This is presumably because of the destruction of the very small grains or large molecules responsible for the PAH features. In ultraluminous infrared galaxies such as Arp 220, the very large amount of dust along our line of sight to the star-forming region also plays a part in this suppression of the PAH features.

The extension of the *ISO* wavelength coverage to 200 microns allowed a better characterization of the cool interstellar dust in galaxies than had been possible with *IRAS*. Most spiral galaxies, whether actively forming stars or not, have a cool, extended dust component with a temperature in the range 15–30K that is caused by emission from interstellar dust illuminated by the stellar radiation field, the infrared 'cirrus'. In many galaxies, the total dust mass estimated by *ISO* was considerably higher than had been estimated by *IRAS*, resolving an apparent discrepancy in the ratio of dust to gas compared with our Galaxy.

At the time of the *ISO* launch there was considerable interest in the idea that much of the dark matter inferred from dynamical arguments to exist in the halos of galaxies might be in the form of low-mass stars or brown dwarfs. Searches were therefore made with *ISO* in the halos of nearby galaxies, especially edge-on spirals where the halo could be imaged close to the centre of the galaxy. No infrared halos were detected, and this more or less ruled out the idea that dark matter halos could be composed of stellar or substellar objects.

The wealth of spectral lines detected by the *ISO* spectrometers provided a powerful insight into the physical conditions in different types of galaxies. Molecular hydrogen, and atomic lines which do not require very high-energy photons to be excited, probe what are known as *photodissociation regions*. These regions are the source of much of the infrared radiation from the interstellar medium in a galaxy and are created when far-ultraviolet radiation from hot stars impinges on dense, neutral, interstellar gas clouds. Small grains or large molecules absorb the radiation and convert it into infrared continuum radiation, spread over a range of infrared wavelengths, or into the broad 'PAH' features. Other spectral lines probe regions of hot gas ionized by massive stars. Combining information from atomic lines and PAH features gives a diagnostic test for whether luminous infrared galaxies are powered by an active galactic nucleus (AGN) or by a burst of star formation (Figure 10.3). For a sample of ultraluminous infrared galaxies, Reinhard Genzel and his co-workers used *ISO* data to show that for 70% of such galaxies the dominant radiation source was a starburst.[9]

Infrared imaging with ISOCAM revealed a major surprise about interacting galaxies such as the Antennae galaxies (Figure 10.4). The beautiful image from the *Hubble Space Telescope* (Plate XIII) shows the dramatic fireworks of star formation, with hundreds of young star clusters triggered by the gravitational interaction between the two galaxies, which are on their way to merging into a single larger galaxy. However, about 80% of the total output from the galaxies emerges in the infrared, and the 15-micron image shows that half of this emission is coming from an optically dark region between the two galaxies,

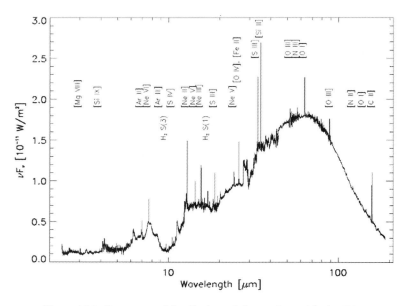

Figure 10.3. Spectrum of the Circinus Galaxy taken with the *ISO* spectrographs. The Circinus Galaxy is mainly a starburst, but with AGN emission features contributing at shorter wavelengths.

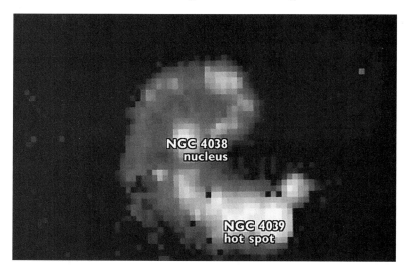

Figure 10.4. *ISO* 15-micron image of the 'Antennae' galaxies, two strongly interacting galaxies in the process of merging together. Comparison with the *Hubble Space Telescope* image of the same galaxies (right panel of Plate XIII) shows that most of the energy in the starburst is hidden from view at optical wavelengths.

where most of the molecular gas and (obscured) star formation are located.[10]

The strongest far-infrared spectral line in galaxies is the 158-micron line of neutral carbon, and the emission in this line is a major contributor to the cooling of the interstellar gas in the galaxies. It is an excellent tracer of star formation in normal and moderately active galaxies. A surprising discovery of *ISO* was that for very active galaxies and ultraluminous infrared galaxies the strength of the line drops, partly because of the greater amounts of dust and gas along our line of sight in these galaxies.

Another feature that turned out to be weaker than expected was the 10-micron silicate feature in Seyfert galaxies. Models of their dust tori tended to predict that Type 1 Seyferts, in which the dust torus is believed to be seen face-on, should show a strong 10-micron emission feature. Before *ISO* the weakness of this feature had been explained as the result of the special geometry of the torus, in which the bulk of its dust is shielded from the ultraviolet radiation of the accretion disk.[11] But these models still predicted a weak silicate emission feature for face-on objects, and this remained a problem. For quasars a controversy developed about their 30–200 micron emission.[12] Dave Sanders argued that the far-infrared emission is direct emission from the AGN, for example from the outer regions of a warped disk.[13] The rival view was that this emission is caused by an associated nuclear starburst distributed on a much larger scale (tens of thousands of light years) than the AGN and its torus (hundreds of light years), and this is the view that seems to have prevailed.

ISO GALAXY SURVEYS

Surveys with *ISO*, mainly with the ISOCAM camera, played an important role in the census of the star-formation history of the universe and in predicting the far-infrared background radiation. With notable exceptions like the redshift 2.3 hyperluminous infrared galaxy F10214+4724, *IRAS* had probed only to about a redshift of 0.3. With its far greater sensitivity, there was a good prospect that *ISO* could probe to redshifts greater than 1. The way to test this was with a series of surveys, and there were over a dozen carried out, most with the broad ISOCAM 15-micron band. Several surveys, to varying depths, were carried out by the ISOCAM instrument team using their guaranteed observing time.[14] Using the open observing time available to all astronomers, a large European consortium carried out the largest *ISO* survey, the European Large-Area *ISO* Survey (ELAIS) of 12 square degrees of sky at wavelengths from 7 to 160 microns. Eighty scientists from 14 European institutions were involved in this survey.[15] We selected areas of the sky with particularly low levels of interstellar dust emission, as indicated by the *IRAS* 100-micron maps, and these areas became the focus for later surveys with the *Spitzer* and *Herschel* space telescopes.

When the *Hubble Space Telescope* carried out its very deep survey in the Hubble Deep Field, we proposed a very deep *ISO* survey in the same area. We found 13 sources and were able to associate them with distant galaxies that had been identified by the *Hubble Space Telescope* (Plate XIV). The infrared power of these galaxies gave us a measure of the star-formation rate in these distant galaxies, and we showed that the rate rises steeply as we look back in time (Figure 10.5).[16] The higher rate of star formation deduced from *ISO* data compared with ground-based or *Hubble Space Telescope* surveys at optical or near-infrared wavelengths was because so much of the starlight in luminous starbursts, over half according to the *COBE* background measurements (Chapter 8), is absorbed by dust and reradiated at infrared wavelengths. Only with *ISO* could we measure the true total energy in starlight in distant galaxies. Catherine Cesarsky's group was able to show that the galaxies found in the deep ISOCAM 15-micron surveys were capable of accounting for the whole of the far-infrared extragalactic background.[17]

While *ISO* was in orbit, two other space missions were flown. The U.S. Air Force launched the MSX mission, which operated in the mid-infrared from 4 to 24 microns, flew for ten months, and devoted about 20% of its observing time to astronomical projects. It surveyed the small strips of the sky missed by *IRAS* and made higher-resolution images of the Galactic plane. The Japanese Space Agency in 1996 launched the IRTS mission, which mapped 7% of the sky across the whole infrared band from 1 to 1000 microns.

THE *SPITZER* SAGA

NASA's *Spitzer Space Telescope* was finally launched on 25 August 2003 after the longest saga of any space astronomy mission. It had been first proposed over 30 years earlier, in 1971, by NASA's Ames Research Center, as a 1-metre class cooled infrared telescope to be flown on the space shuttle, at a time when the shuttle was expected to offer frequent flight opportunities and missions lasting up to 30 days.[18] It became known as the *Shuttle Infrared Telescope Facility* (*SIRTF*) and continued to be studied at Ames throughout the 1970s and 1980s. As I discussed in Chapter 7, in

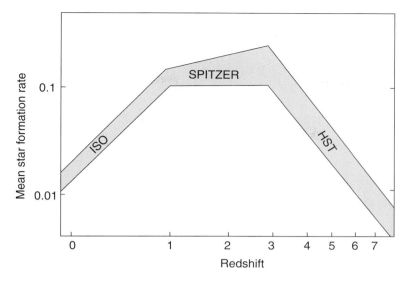

Figure 10.5. Schematic history of the average star-formation rate per unit volume as a function of redshift, showing that the peak epoch for star formation was at a redshift of about 2, or about 3 billion years after the Big Bang. Surveys with *ISO* established the trend from redshift 0 to 1, surveys with *Spitzer* extended this to redshift 3, and the behaviour at redshifts 3–7 has been established using surveys with the *Hubble Space Telescope* and large ground-based telescopes.

1976 NASA decided to accept the recommendation of leading U.S. infrared astronomers to first go for a smaller and simpler survey mission, which became *IRAS* and flew in 1983. In May 1983, four months after the launch of *IRAS*, NASA issued an Announcement of Opportunity (AO) for focal-plane instruments for *SIRTF*, 'an attached Shuttle mission with an evolving scientific payload', but hinted at 'a probable transition to a more extended mode of operation, possibly in association with a future space platform or space station'. The first shuttle flight for *SIRTF* was projected to be in about 1990, with the second flight approximately one year later. The AO suggested that one or two instrument specialists might be needed to fly on the shuttle. However, the immense success of the free-flying *IRAS* mission, coupled with concerns about the cleanliness of the shuttle environment in terms of particles and radiation, changed NASA's thinking, and by the time the successful instrument bids were announced in June 1984, *SIRTF* was envisaged as a free-flying mission with the same orbit as *IRAS*. The three instruments selected were short- and long-wavelength infrared-array cameras, IRAC and

Figure 10.6. The *Spitzer* (then *SIRTF*) Science Working Group in 1984. Back row (from L): George Newton, Dan Gezari, Ned Wright, Mike Jura, Mike Werner, Fred Witteborn; front row: Giovanni Fazio, George Rieke, Nancy Boggess, Jim Houck, Frank Low, Terry Herter.

MIPS, led by Giovanni Fazio[19] and George Rieke,[20] and an infrared spectrograph, IRS, led by Jim Houck.[21] Frank Low became facility scientist, and Mike Werner became the project scientist and, over the next 19 years until eventual launch, the tireless spokesperson for the project (Figure 10.6). In 1985, *SIRTF* was identified as one of NASA's four 'Great Observatories', along with the Compton gamma-ray observatory, the Chandra x-ray observatory and the *Hubble Space Telescope*. From 1987 it began to be realized that a high Earth orbit, at 100,000 kilometres from Earth rather than the 900 kilometres previously envisaged, would have major thermal and operational advantages. *IRAS* veterans Gerry Neugebauer and Fred Gillett joined Frank Low in successfully pressing this view at a 1989 NASA review. *SIRTF* was now envisaged as a 5.7-ton mission, carrying 3800 litres of liquid helium to ensure a 5-year life, to be launched by the massive Titan-Centaur rocket (Figure 10.7). *SIRTF* was selected as the highest-priority new initiative for the 1990s by the US Decadal Review, chaired by John Bahcall.

At this point *SIRTF* entered an uncertain state for several years. Problems with the *Hubble Space Telescope* and *Galileo* missions meant that funding and effort were shifted away from *SIRTF*. Dan Golden took over the leadership of NASA with the mantra 'faster, cheaper, better', and *SIRTF* had to endure the first of its 'descopes', reductions in the cost and scope of the mission. In 1993, it became a 2.5-ton mission carrying 920 litres of helium for a 3-year life, to be launched on the medium-sized Atlas rocket. But this was swept aside by a NASA ruling that no space mission could exceed $500 million in cost. NASA's new astrophysics director, Dan Weedman, declared according to Mike Werner[22] that 'we could use any launch vehicle we liked, as long as it was a Delta'. The Delta is the United States' workhorse rocket for launching small satellites and had launched *IRAS* and *COBE*. A further radical descope would be necessary.

Two years earlier, in 1991, Harley Thronson of the University of Wyoming had proposed to NASA an entirely new concept for an infrared space mission, *Edison*, with a passively cooled telescope. The idea, originally developed by British astronomer Tim Hawarden for a mission proposal to ESA in 1989, is that the telescope orbits as far as possible from Earth, with excellent thermal shielding, and is allowed to cool to its ambient temperature. Cryogen is then needed only to cool the instrument package. The whole spacecraft can then be much lighter, with the telescope being launched warm. The *SIRTF* team realized that this idea was the only hope for keeping their mission within cost and of sufficiently low mass to be capable of being launched by a Delta. At a brainstorming meeting in November 1993, Frank Low sketched out a warm-launch architecture for the spacecraft. *SIRTF* now began to take its final shape. It would be an 85-centimetre telescope, only very slightly less than the 90-centimetre size of the Titan-Centaur monster, but now there would only be 350 litres of liquid helium, one-tenth that originally proposed. The mass was down to 850 kilograms, one-sixth that originally planned. And the total cost was down by a similar factor. Amazingly the science performance, at least in terms of the mirror size and lifetime of the mission, now projected to be at least 5 years, were undiminished compared with the original design. However, the warm mirror would make longer-wavelength observations more difficult. And more seriously, the descope to the satellite was accompanied by a descope of the instrument package. It was unfortunate that there was only one imaging band between 8 and 70 microns, at 24 microns, that there was no high-resolution spectrograph, and that the IRS operated only to 35 microns. These compromises were compensated by the exceptional sensitivity of the mid-infrared camera (IRAC) at 3.6–8 microns,

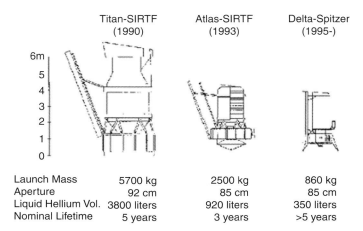

	Titan-SIRTF (1990)	Atlas-SIRTF (1993)	Delta-Spitzer (1995-)
Launch Mass	5700 kg	2500 kg	860 kg
Aperture	92 cm	85 cm	85 cm
Liquid Hellium Vol.	3800 liters	920 liters	350 liters
Nominal Lifetime	5 years	3 years	>5 years

Figure 10.7. The evolution of the *SIRTF* mission, finally launched as *Spitzer* in 2003.

of the far-infrared camera (MIPS) at 24 microns, and of the spectrograph (LRS) at 5–35 microns. The IRAC array had 256×256 pixels, compared with the 32×32 of *ISO*'s ISOCAM array, so it could map a much larger area of sky at any one time. Thronson's *Edison* mission was never approved, but its revolutionary passive-cooling concept saved the *SIRTF* mission and was to be central to the European Space Agency's *Herschel* mission and to the international James Webb Telescope mission (Chapter 12).

The key to the passive-cooling strategy for *SIRTF* was the orbit, which needed to take the satellite as far away from the Earth as possible. This was achieved by launching *SIRTF* into an orbit trailing behind the Earth. Every day, as it gets farther and farther away from the Earth, the satellite points its high-gain antenna towards the Earth and transmits the past day's data. The spacecraft is oriented so that the solar panels point to the Sun, the back side of the outer shell radiates to space, and the telescope observes away from the Sun and Earth. *SIRTF* was finally launched in 2003 and was renamed the *Spitzer Space Telescope*, in honour of the American astrophysicist Lyman Spitzer (Figure 10.8). Its coolant lasted until 16 May 2009, coincidentally just two days after the launch of *Herschel*. The instrument package has therefore now warmed up, but the 3.6- and 4.5-micron detectors can still function and *Spitzer* continues to operate at these wavelengths, carrying out deep surveys and studying exoplanets. About ten years after launch, around 2013, *Spitzer* will pass out of communications reach of the Earth.

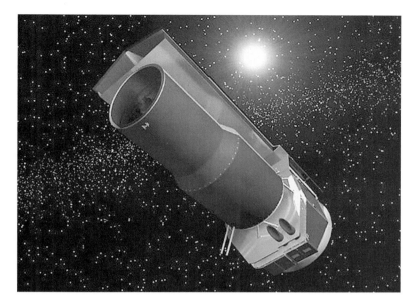

Figure 10.8. The *Spitzer Space Telescope*, launched in 2003.

EXTRAGALACTIC SCIENCE WITH *SPITZER*

Much of the first year of observation with *Spitzer* was devoted to six Legacy Surveys: the 'SWIRE' extragalactic survey of 49 square degrees of sky (about 250 times the area subtended by the Moon); the 'GOODS' very deep extragalactic survey of a much smaller area, one-fifth the area of the Moon, including the Hubble Deep Field; the 'SINGS' survey of nearby galaxies; the 'GLIMPSE' survey of the Galactic plane; the 'c2d' survey of star-forming regions; and the 'FEPS' survey of dust debris disks.[23] The emphasis on Legacy Surveys continued in subsequent years, and altogether over 30 such surveys were carried out, covering a wide range of science. Most of these Legacy Surveys involved collecting supplementary information at noninfrared wavelengths and so they became multiwavelength surveys, covering much of the electromagnetic spectrum. For extragalactic surveys, determining the redshift of the galaxies was crucial, so that the distance of the galaxy was known and its luminosity could be determined. However, in most cases it was impractical to obtain spectroscopic redshifts for all the galaxies. The SWIRE survey, for example, contained over a million galaxies. It was therefore necessary to try to estimate the redshift by modelling the optical energy distribution. This was known as the 'photometric redshift' method. Some of the most wonderful images from *Spitzer* have come from the SINGS survey of nearby galaxies, and they are as spectacular as those of the *Hubble Space Telescope* (Plates XV to XVIII). Detailed analysis of these images, especially when combined with optical and ultraviolet data, has led to a better understanding of the physics of star formation.

Two teams used the GOODS survey data to estimate the star-formation history of galaxies.[24] The star-formation rate in galaxies was estimated using the 24-micron luminosity, because for starburst galaxies there is a good correlation between the 24-micron luminosity and the total far-infrared luminosity, which in turn is a good measure of the star-formation rate. The teams confirmed the result from the *ISO* surveys that the star-formation rate increases steeply, by a factor of 10, between the present epoch and redshift 1, eight billion years ago. The *Spitzer* surveys extended to higher redshifts than the *ISO* surveys and found that the star-formation rate is level from redshift 1 to redshift 3, 11 billion years ago, and then perhaps declines at higher redshifts (see Figure 10.5). The increase in star-formation rate with redshift was steepest for the most infrared-luminous galaxies, suggesting that these galaxies were more dominant at high redshifts. The *Spitzer* 3.6–8-micron data allowed the stellar masses and ages of galaxies at high redshifts to be determined with much higher precision than before. It was realized that the buildup in stellar mass in the most massive galaxies must have been nearly completed by five billion years after the Big Bang.[25] The obscured star formation detected at infrared wavelengths accounts for about 75% of the total star formation in the universe.

Perhaps the most surprising *Spitzer* discovery was the detection of some of the very highest-redshift galaxies known. Eeichi Egami and his collaborators combined *Spitzer* and *Hubble Space Telescope* observations to estimate the redshift for a galaxy gravitationally lensed by a cluster to be in the range 6.6–6.8 (Figure 10.9).[26] It seems remarkable that this very distant galaxy could be detected with a telescope with a mirror only 0.85 metres in diameter. This galaxy was being seen only one billion years after the Big Bang. The researchers also found that the stars in the galaxy were surprisingly old, 50–450 million years. Several groups studied galaxies at redshifts 5–6 and estimated star-formation rates and stellar masses. As much as a few per cent of the stellar mass in galaxies was in place by redshift 5, only 1.3 billion years after the Big Bang. However, these early stars do not seem to provide sufficient ultraviolet radiation to explain why most of the intergalactic gas at this epoch is already ionized. To do this there must be stars and

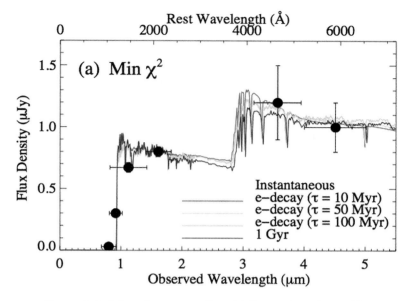

Figure 10.9. Infrared spectrum of a redshift 6.7 galaxy detected by *Spitzer* and the *Hubble Space Telescope*. The filled circles with error bars are the detections at near-infrared wavelengths by the *Hubble Space Telescope* and at 3.6 and 4.5 microns by *Spitzer*. The curves are different models for young galaxies (Egami et al. 2005). The galaxy has been gravitationally lensed by the cluster Abell 2218 (see Plate VI).

galaxies that are at even higher redshifts, and to find them a new generation of telescopes will be needed (see Chapter 12).

Identification of high-redshift galaxies depends on accurate predictions of the spectral energy distributions of these young galaxies. Models for optical and near-infrared emission from galaxies were greatly strengthened by the appearance of new syntheses of the spectra of starlight in galaxies under different possible assumptions about their star-formation histories.[27]

There are two ways of studying the evolution of the galaxy population out to high redshifts. The first is to count the galaxies as a function of brightness at each wavelength and model these counts in terms of the populations we know about locally. The second is to add up the total radiation detected from all galaxies, the integrated galaxy light, and model the spectrum of this radiation in terms of known galaxy populations. The *Spitzer* mission made great progress in both of these areas. The *Spitzer* galaxy counts at 24, 70 and

160 microns reached sufficiently deeply for the evolution of infrared galaxies out to high redshifts to be modelled. The story is similar to that found from studying the 24-micron luminosities in different redshift ranges: a very strong increase in galaxy luminosity as we look back in time to redshift 1, a flattening of the evolution from redshifts 1 to 3, and then a decline at higher redshifts.[28] The infrared background radiation at 140 microns detected by *COBE* could be accounted for by summing the sources directly observed in deep *Spitzer* surveys.[29] This means that the *Spitzer* surveys give us an almost complete picture of the infrared sky and of the evolution of the infrared galaxy population.

Spitzer's very sensitive IRAC bands at 3.6, 4.5, 5.8 and 8 microns were ideal for detecting the dust tori around active galactic nuclei.[30] We modelled over 200,000 galaxies in our very large SWIRE survey in terms of a simple set of infrared templates corresponding to quiescent, 'cirrus' galaxies, normal and extreme starbursts, and AGN dust tori, and found that almost one-third of SWIRE galaxies contained a dust torus component, indicating the presence of an AGN.[31] Several studies set out to compare the *Spitzer* picture of the AGN population with that from x-rays, mainly through surveys with the Chandra x-ray satellite. All agree that there is a significant population of AGNs that are so heavily shrouded in gas and dust that much of the x-rays, especially at 'softer' x-ray wavelengths, are absorbed. However, there is still rather wide disagreement about what fraction of AGNs fall in this heavily shrouded class.

The high sensitivity of the IRS spectrometer yielded many interesting results on the nature of star formation in galaxies. Crystalline silicates were detected in the spectra of ultraluminous infrared galaxies, which suggests some special phase in the formation of these objects (Figure 10.10).[32] The fraction of silicates that are crystalline rather than amorphous, though small, is far higher in ultraluminous infrared galaxies than in the interstellar medium of our own Galaxy, and this suggest that the timescale for their injection is very short, pointing to massive stars as their origin. Galaxies found at high redshifts have PAH feature strengths similar to normal local galaxies and so must already have interstellar gas and dust with heavy-element abundances close to that in the Galaxy (i.e., they have already made most of their heavy elements). The detection of PAH features in galaxies at redshift 3 shows that carbon-rich materials must already have been in place only two billion years after the Big Bang, so galaxy formation must have taken place at much higher redshifts and earlier times.

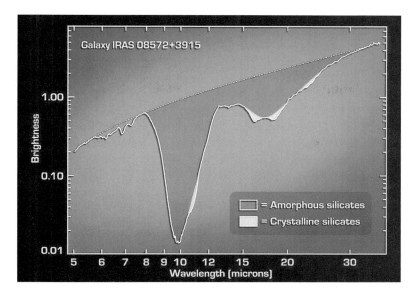

Figure 10.10. The surprising detection by Spitzer of crystalline silicates in the spectra of ultraluminous infrared galaxies. Superposed on the usual deep absorptions at 10 and 18 microns caused by amorphous silicates, there are small features resulting from crystalline silicates.

A weak silicate emission feature was found in the spectra of some quasars, and this resolved a puzzle for unified models of quasars in which different types of active galactic nuclei were simply a reflection of different viewing angles to the nucleus and its surrounding dust torus.[33] Hitherto the silicate feature had always been seen in absorption, whereas models of the dust tori predicted that quasars in which the dust torus is seen face-on should have at least weak silicate emission.

GALACTIC AND PLANETARY SYSTEM SCIENCE WITH *SPITZER*

Spitzer made the first direct detection of an extrasolar planet, characterized planet-forming and planetary debris disks around solar-type stars, and showed that substellar objects with masses less than ten times that of Jupiter form through the same processes as solar-mass stars.[34]

As we have seen, stars are formed deep within dense clouds of molecular gas and dust. They are hidden from view at optical wavelengths because of extinction by dust but peek out at infrared and submillimetre wavelengths. When a young star switches on and starts to drive away the surrounding gas and dust, it is common for the disk of accreted material from which the star accumulated to remain behind. If it is dense and massive enough to support the formation of planets, it is called a *protoplanetary* disk. In well-studied nearby clouds like those in Orion and Taurus, these disks tend to have masses in the range one-thousandth to one-hundredth of the mass of the Sun and sizes of about 100 astronomical units, similar to the farthest limit of the cometary cloud in the Solar System. For comparison, the total mass of planets, asteroids, comets and other objects in the Solar System is just over one-thousandth of the mass of the Sun. Dust in protoplanetary disks aggregates together quite quickly, and within ten million years objects with sizes up to 100 kilometres form. These become the building blocks for planets and are known as *planetesimals*. From these planetesimals, the solid cores of planets accumulate through collisions and can then accrete gaseous mantles if they are not too close to the parent star. At this stage, we call the residual material not accumulated into planets a *planetary debris disk*.

Huge numbers of young stellar objects (YSOs) were found by *Spitzer* in large-area surveys of known regions of star formation in the Galaxy, and the infrared colours were then used to identify the different types of YSOs and investigate their evolution. More than half of YSOs have disks. The sensitivity of the *Spitzer* Infrared Spectrograph meant that spectra could be taken of fainter objects than with *ISO*, for example the very young Type O objects. These were found to contain very low-luminosity infrared cores. The mid-infrared spectra showed strong absorption features of water and carbon dioxide ice believed to accumulate just before the dust settles to the disk and is incorporated into planetesimals (Plate XIX). Individual massive protostars, from ten solar masses upwards, could be identified as far away as our neighbouring galaxy, the Large Magellanic Cloud.

PLANETARY SYSTEM FORMATION AND EVOLUTION

Long after the dissipation of a protoplanetary disk, collisions of asteroids or evaporation of comets around a star can inject dust into its circumstellar environment. The tenuous disk that forms and regenerates through this process is called a planetary debris disk. In the Solar System, this corresponds to the zodiacal dust cloud and asteroid belt. Such disks around other stars can best be studied through their infrared emission. The asteroid belt and zodiacal dust cloud contribute only one-ten-millionth of the luminosity of the Solar System, so it was a surprise when *IRAS* detected debris disks contributing fractions of their star's luminosity a thousand times greater than this. Even so, only a very small amount of dust is needed to explain such debris disks: less than one-hundredth of an Earth mass. *IRAS* and *ISO* demonstrated that debris disks contributing more than 0.001% of the total light from the star occur around 15% of nearby stars like the Sun. These missions also established a rough timescale of a few hundred million years for the decay of the systems around young stars, although some older stars still have large excesses. *Spitzer* was able to detect dozens of new debris disks.[35]

The spectral energy distributions of the great majority of debris disks are remarkably similar and indicate that the radiating dust is at a temperature of about 70 K. This means that the material is at a distance of about 10–100 astronomical units (1 AU is the mean distance of the Earth from the Sun) and so is analogous to the region in the Solar System known as the Kuiper Belt. For old stars similar to the Sun, the *Spitzer* spectra show no radiation at wavelengths less than 33 microns, indicating that these rings of debris have little material in the region occupied by the terrestial planets (Mercury, Venus, Earth and Mars) in the Solar System. For hotter stars such as Vega, the material does extend farther in, which explains how the *IRAS* team was able to detect the Vega disk. There are a few stars with excess emission only at 24 microns, and this may result from dust generated by collisions between asteroids. Imaging of debris disks with *Spitzer*, with submillimetre telescopes, and in the optical, either with ground-based telescopes or the *Hubble Space Telescope*, shows a variety of structures. Generally there is a ring of dust at around 100 AU, typically of dust grains about 10 microns or smaller in size. Sometimes this dust has migrated farther inwards. In the case of Vega, the debris has been traced out to 1000 AU and photon pressure from the star is believed to be driving grains outwards. The debris is believed to be fairly transient, generated by collisions between planetesimals.

Debris disk studies with *ISO* were the first to benefit from accurate stellar dating,[36] which allowed the disks to be placed in an evolutionary sequence so that the gradual falloff of dust mass with time, especially over the first ten million years, can be traced. The increased sensitivity of *Spitzer* allowed much weaker debris disks to be detected and all but bridged the gap between the very dusty systems first detected by *IRAS* and the much lower levels of dust and debris in the Solar System. An important distinction between debris disks and protoplanetary disks is that the former contribute less than 1% of the total stellar luminosity, whereas the latter have higher fractional luminosities. Debris disks with fractional luminosities above 0.1% are invariably young systems less than a hundred million years old.

PROTOPLANETARY DISK EVOLUTION, END OF PROTOPLANETARY DISK PHASE

The infrared emission from debris disks is coming from dust grains with sizes in the range from 0.1 microns to 1 millimetre, but there must also be a belt of much larger objects, planetesimals, together with a mechanism for turning them into dust. Planet formation is a dynamic process that continues for the first billion years of a star's life. Observations of debris disks can set constraints on the planet-formation process, and some debris disks may be remnants of protoplanetary disks.[37]

Most stars are born with protoplanetary disks, in which the planetesimals needed both for formation of planets and to generate debris disks are formed. Initially most of the protoplanetary disk is gas, with just 1% in the form of submicron-sized grains. Protoplanetary disks have a broad range of masses, with an average of about 30 Earth masses. The outer disk radius is in the range from 10 to 1000 AU. The fraction of Sun-like stars with protoplanetary disks declines sharply with age from 100% at zero age to zero at ten million years, with a typical lifetime of 1–6 million years. The disks appear to be cleared from the inside out, partly by grains sticking together and falling into the star and partly by the formation of planets. Stars with disk dust masses less than one Earth mass are all older than 30 million years, and this gives a clear indication that these are debris disks, which need to be continuously replenished.

Once kilometre-sized planetesimals have formed, there is a period of runaway growth, and then slower growth to build up planetary cores 2000–3000 kilometres in size, with frequent collisions resulting either in mergers or in ejection of one of the components. The larger cores stir up the surrounding planetesimals, and these then tend to collide with each other destructively, generating dust, which is driven out of the system by the pressure of radiation from the star. At the end of the protoplanetary disk phase, a newly formed star is expected to be surrounded by one or all of the following: a planetary system, with planets ranging in size from Mercury to Jupiter, a remnant of the protoplanetary disk, a planetesimal belt in which planets continue to grow, or a planetesimal belt that is being ground down to dust, like the Solar System's asteroid belt.

DEBRIS DISK EVOLUTION

Although collisions between the largest planetesimals in a planetesimal belt are rare, they can release large amounts of dust, which is ground to smaller size by collisions. However, to maintain the observed debris disks, frequent collisions involving smaller planetesimals are needed. In the Solar System, we can see the debris from individual asteroid collisions in the form of the dust bands stretching around the zodiac, which were discovered by *IRAS*. The collisional cascade required to understand the much denser debris disks seen around some other stars may be triggered by gravitational stirring by protoplanets, of radius 2000 kilometres, once these have formed. Seeing a sharply defined debris ring may tell us that a planet has formed inside the inner edge.[38]

Spitzer 24- and 70-micron observations have proved to be a powerful tool in probing debris disks. The 24-micron excess is seen in debris disks around stars up to 100 million years old, and this is consistent with the timescale for terrestrial planet formation, though it is also possible that we are witnessing the grinding away of asteroid belts associated with these stars. The 70-micron excesses from debris disks are seen over much longer timescales, up to 10 billion years, and this suggests steady-state evolution of these much more extended (> 30 AU) disks.

Direct imaging of debris disks shows that several different processes are at work, including steady-state evolution, gravitational stirring by protoplanets, and the likely presence of planets at the inner edge of truncated disks. There is a surprisingly high incidence of warps and asymmetries, which points to the distorting presence of planets.

From the crater record on the Moon, we know that the Earth started its life in a dangerous environment. The heavy bombardment of the Earth and Moon ended about 700 million years after the formation of the Solar System. Throughout the period of bombardment there would have been continuous production of debris, with spikes in grain production whenever a particularly large collision occurred. These may be connected to dynamical instabilities triggered by orbital migration of the giant planets.[39] The evolution of planetary debris disks with time may therefore trace the dynamical evolution of their host planetary systems.

EXTERNAL PLANET OCCULTATIONS

One particularly exciting result from *Spitzer* was the direct detection of an exoplanet through occultation by its parent star. The first confirmed exoplanet orbiting a star like the Sun was the 51 Pegasi system,[40] which was found in 1995 through the very small oscillations of the parent star induced by the gravitational pull of its planet. This was the first example of a 'hot Jupiter', a planet with a mass comparable to Jupiter but located extremely close to its parent star, at only 0.05 AU, about the distance from Mercury to the Sun. Hundreds of such systems are now known.

To detect such planets through their infrared radiation, we need a system that is edge-on to us so that we can see the planet passing in front of and behind its parent star. Two groups using *Spitzer* in 2005 managed to detect the small dimming in infrared light (~0.2%) when the planet around the star HD 209458 passed behind its parent star.[41] Subsequently *Spitzer* has been used to characterize dozens of exoplanets via their transits across their parent star (Plate XX).

If there was a disappointment with *Spitzer*, it was that, like *ISO*, the performance at long wavelengths, the 70- and 160-micron channels of MIPs, was not as good as had been hoped. It was also frustrating that, as a result of the descoping of the mission back in 1993, there were no photometric bands between 8 and 24 microns. This was remedied in the IRC mid-infrared instrument on the Japanese *Akari* mission,[42] launched on 22 February 2006. The IRC instrument has a total of nine bands from 2 to 25 microns and can be expected to make interesting diagnostic analyses of the PAH features both in local star-forming clouds and in distant galaxies. The main goal of *Akari* was to carry out the first all-sky far-infrared survey of the sky since *IRAS*, at wavelengths from 70 to 140 microns. It had been hoped that it would reach to a sensitivity at least ten times deeper than *IRAS*, but a combination of less sensitive detectors than hoped and insufficient redundancy in the survey scanning resulted in the survey not achieving this goal. However, the smaller far-infrared detectors on *Akari* gave it much greater spatial resolution than *IRAS*, and this paid off in much improved images of the Galactic plane and other bright areas of the sky such as the Orion, Taurus and Ophiuchus molecular clouds and the Large Magellanic Cloud.

The achievements of *Spitzer* across the whole range of astrophysics have been remarkable. Although the helium coolant ran out on 16 May 2009, *Spitzer* will be able to operate for several more years as a warm mission at wavelengths of 3.6 and 4.5 microns. Several deep surveys will be carried out, and because these wavelengths are critical for the identification of very high-redshift galaxies, we can still expect some exciting discoveries. Characterization of exoplanets is also one of the main science goals of the *Spitzer* warm mission. In the next chapter, I pull together everything we know about our dusty solar system and its complex and dynamically fascinating debris disk.

11

Our Solar System's Dusty Debris Disk and the Search for Exoplanets

Apart from the Sun, everything in the Solar System emits the bulk of its energy in the infrared. Of course, in reflected optical light from the Sun we do see the planets, comets, zodiacal dust and asteroids, and much of what we have learned about the Solar System has come from optical telescopes. But to see the full picture and make the connection with other planetary systems, we need the infrared vision of the Solar System. We have seen that the infrared space missions *IRAS*, *ISO* and *Spitzer* made a series of discoveries about the Solar System and other planetary systems. In particular, the discovery by *IRAS* of dust debris disks opened the way for the exploration of other planetary systems in the process of formation. In this chapter, we focus on the debris disk in our own solar system: the asteroid belt, the zodiacal cloud, the Kuiper Belt, and comets and Centaurs. This leads on naturally to the exciting search for planets around other stars and ultimately for planets like Earth which could sustain life.

The five visible planets – Mercury, Venus, Mars, Jupiter and Saturn – have been known and studied since antiquity. In 1543, Nicolaus Copernicus (1473–1543) gave us our modern picture of the planets, including the Earth, orbiting the Sun in approximately a common plane, the ecliptic plane, with the Moon orbiting the Earth. In 1605, Johannes Kepler (1571–1630) showed that the orbits of the planets were approximately ellipses, and Newton in 1687 explained these orbits in terms of the inverse-square law of gravitation. The seventeenth century also saw the discoveries of moons around Jupiter by Galileo in 1610 and of Saturn's moon Titan by Huygens in 1655. As we saw in Chapter 2, William Herschel discovered Uranus in 1781, and the eighth planet, Neptune, was discovered by Urbain Leverrier

I have made extensive use of the reviews by Luu and Jewett (2002) and Wyatt (2008).

(1811–1877) and Jean-Couch Adams (1819–1892) in 1846. The nineteenth century also saw the discovery of asteroids in orbit between Mars and Jupiter and the discovery of the moons Phobos and Deimos around Mars.

The Solar System contains not just its eight planets but also vast numbers of smaller bodies, ranging from objects more than 2000 kilometres in diameter, such as Pluto and the asteroid Eris, down to tiny, submicron-sized dust grains. By analogy with the systems around other stars, we can call everything in the Solar System, apart from the eight planets and their moons, its debris disk. The bulk of this debris disk is concentrated in two belts, the asteroid belt at 2–3.5 astronomical units (AU) and the Kuiper Belt at 30–50 AU. There has been a long ongoing debate about where the dividing line is between a planet and a large member of the debris disk, and this was resolved, for the moment at least, by the declaration of the 2006 General Assembly of the International Astronomical Union in Prague that the largest members of the asteroid and Kuiper belts, including the former planet Pluto, should be called dwarf planets (see Plate XXI). The adopted definition of a planet is that it must be an object with sufficient mass that its self-gravity imposes a nearly round shape and it must also be dominant enough to clear the neighbourhood around its orbit. It was the second condition that ruled out the larger asteroids, and Pluto and other trans-Neptunic bodies, as planets. The adopted definition followed a turbulent discussion over the two weeks of the General Assembly during which it appeared that the outcome could go either way.[1]

In addition to the asteroid and Kuiper belts there is an extended disk of dust stretching from the Sun out to beyond the asteroid belt called the zodiacal dust cloud. This is responsible for the phenomenon of the zodiacal light, which can be seen from very clear, dark sites, especially near the tropics (Figure 11.1). The zodiacal light has sometimes been referred to as the 'false dawn' and may be mentioned in the Rubaiyat of Omar Khayyam, the great Persian astronomer-poet of the twelfth century:

> When False dawn streaks the East with cold, gray line
> Pour in your cups the pure Blood of the Vine

The discovery of the zodiacal light in the West is attributed to the French astronomer Domenico Cassini (1625–1712), who described it in 1683 and put forward the correct explanation, that it is sunlight reflected from small grains of dust. The spatial distribution of zodiacal dust began to be modelled in detail during the 1950s and 1960s, with direct information on Solar System dust beginning to come from early

Figure 11.1. Photo of zodiacal emission taken from Tenerife by Brian May in 1971.

space missions. In the late 1960s and early 1970s, the detailed motion of the zodiacal dust began to be studied by Jim Ring's Imperial College group. With the launch of *IRAS* in 1983, it became possible to map the infrared emission from zodiacal dust across the whole sky[2] and model its distribution in great detail. This was done by Stan Dermott and his collaborators at Cornell University[3] and by my group at Queen Mary College.[4] An enjoyable footnote to my involvement with zodiacal dust was that Brian May, who had been doing a PhD on the kinematics of zodiacal dust in the Imperial College group in the 1970s when his rock band, Queen, took over his life, decided to return to Imperial in 2005 and complete his PhD under my supervision. It was a great achievement for Brian to complete his thesis after a gap of thirty years, and it certainly brought some excitement into my life.

The final constituent of the Solar System is the Oort cometary cloud. In 1950, Jan Oort made a study of the orbits of all the known comets and concluded that they must have had their origin in a cloud of comets situated thousands of times farther from the Sun than the Earth, reviving an idea from 20 years earlier by the Estonian astronomer Ernst Opik. Today we think the Oort Cloud stretches from 2000 to 50,000 AU or beyond, with an inner doughnut-shaped cloud stretching from 2000 to 20,000 AU and an outer spherical cloud stretching from

20,000 to 50,000 AU (Figure 11.2). The Cloud contains perhaps a billion comets larger than 20 kilometres in size and many more smaller ones. The total mass of material in the Oort Cloud probably amounts to only five times the mass of the Earth. Comets represent primordial material left over from the birth of the Solar System, and most orbit out there without ever troubling the inner Solar System. Occasionally gravitational deflection resulting from a passing star stirs up the Cloud and sends a comet on a path deep in towards the Sun. Some actually hit the Sun, and many such encounters have been imaged by the joint ESA and NASA Solar and Heliospheric Observatory (SOHO). Comets are aggregates of dust and ice, some of which they shed along their orbits. The trails of dust that they leave along their orbits were first detected by *IRAS* in 1983 and have been spectacularly imaged in infrared light by *Spitzer* (Plate XXII).

The Solar System's debris disk has not always looked as it does today. Both the asteroid and Kuiper belts are believed to have started off with much more mass than they have now. The history of the Solar System has been punctuated by violent events that changed the local debris population, such as collisions between protoplanets or asteroids. Evidence for this comes from a variety of sources: the orbits of the planets, the size distribution and dynamical structure of the debris belts, the cratering records and compositions of planets and debris, and deep ocean sediments on the Earth. However, the interpretation of this evidence is still hotly debated, particularly for the era of the first billion years after the Solar System formed, when the planetary system was undergoing its last stages of accretion and settling towards its current configuration.

Although the Solar System planet orbits have been stable for billions of years, the evolution of the terrestrial planets, asteroids and Kuiper Belt objects may be chaotic both on longer and shorter timescales.[5] The dynamical structure of its debris disk, and the cratering records of the terrestrial planets and the Solar System's moons, suggest there has been significant restructuring during the Solar System's evolution. For example, the dynamical structure of the Kuiper Belt provides evidence of effects of the migration of Neptune. Age dating of lunar samples shows that the cratering rate on the Moon was much higher in the first 600–800 million years than subsequently. This cataclysmic period about 700 million years after the Solar System formed is called the Late Heavy Bombardment and is thought to have been caused by a high flux of planetesimals originating in the asteroid belt.[6] To explain these various observations it has been suggested that

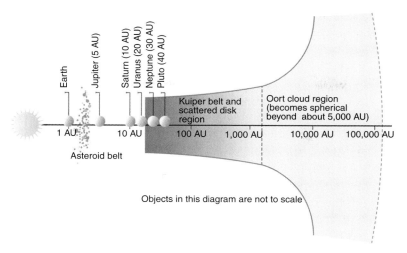

Figure 11.2.Schematic profile of the Solar System, showing the planets, asteroid belt, Kuiper Belt and Oort Cloud.

Uranus and Neptune originally formed between Jupiter and Saturn, or that there were other planets like Uranus and Neptune that have been lost, and that the planetary system was pushed into instability by interaction with the remnant planetesimal disk. Chaotic dynamical evolution of planetary systems can also explain the high eccentricities seen in many extrasolar planets.

One of the goals of studying debris disk evolution around other stars is to try to place the Solar System in context. If we compare the nearest Sun-like stars, then we can see that the Solar System is not particularly unusual. It is certainly not exceptionally dusty, because the nearby stars Epsilon Eridani and Tau Ceti both have more than 20 times as much dust as the present Kuiper Belt. Likewise we cannot yet say that the Solar System is exceptionally low in dust, because a debris disk like our Kuiper Belt around even a nearby star is not yet detectable. Currently the Kuiper Belt accounts for one-ten-millionth of the luminosity of the Solar System, contains between one-thirtieth and one-tenth the mass of the Earth in material, and is located about 40 astronomical units from the Sun. It has been suggested that the Kuiper Belt underwent a dynamical instability about 800 million years after the formation of the Solar System that would have greatly depleted its mass.[7] Before this it could have contained 30 Earth masses of material at a distance of 25 AU. Such a system around a nearby star would result in clearly detectable emission at 70 microns, so the Solar System's debris disk might have been very typical of those studied with *Spitzer* at this wavelength, though unusual in having a dynamical instability that depleted its disk.

KUIPER BELT AND TRANS-NEPTUNIAN OBJECTS

The Kuiper Belt consists of a large number of small, solid bodies in heliocentric orbit beyond Neptune. Only definitely identified in 1992, it holds the key to understanding the early Solar System, as well as the origin of outer Solar System objects such as short-period comets and the Pluto-Charon system. Kuiper Belt objects are believed to be ancient relics of the Sun's accretion disk, from which the Sun grew and its planets formed.[8]

The first object detected beyond Neptune was Pluto, found in a photographic survey by Clyde Tombaugh in 1930. At the time it was called the ninth planet, but it turned out to have a mass of only 0.002 Earth masses, six times lighter than the Moon. In 1951, the Dutch American astronomer Gerard Kuiper discussed the idea of a ring of small bodies beyond Pluto, which would be scattered to the Oort cometary cloud by Pluto.[9] Comets would then be returned to the inner solar system through gravitational perturbations by passing stars.

The first Kuiper Belt object (KBO) was detected in 1992 by Jane Luu and Dave Jewett,[10] and within 10 years over 400 were known, with over 1000 now known. The known KBOs have been classified into three groups, according to their orbits: the classical KBOs, the resonant KBOs and the scattered KBOs. Their orbits, like those of all objects that remain within the Solar System, are ellipses, with the Sun as one focus. The orbit can be characterized by three numbers: the maximum distance from the Sun (semimajor axis), how non-circular the orbit is (the eccentricity), and the tilt between the plane of the body's orbit and the ecliptic plane (the inclination). The classical KBOs have orbits whose most distant point from the Sun lies in the range of 42–48 AU and have small eccentricities (nearly circular orbits). At this distance they are relatively immune to the effects of Neptune, at 30 AU, which defines the inner edge of the Kuiper Belt. The average inclination of the classical KBOs is a measure of the thickness of the Kuiper Belt, and the distribution of inclinations is a measure of how dynamically excited the belt is. The Kuiper Belt is found to be quite thick, with a mean inclination of 20 degrees, much thicker than the disk from which the KBOs are believed to have formed. The classical Kuiper Belt is estimated to contain 10 objects with radii larger than 1000 kilometres, 30,000 objects larger than 50 kilometres and ten billion objects larger than 1 kilometre. For comparison, the radius of Pluto is only 1150 kilometres, so there are likely to be several KBOs bigger than Pluto. You can see why the International Astronomical Union felt obliged to reconsider the definition of a planet and downgrade Pluto to the status of a dwarf planet. The reason for the rather sharp outer edge to the classical Kuiper Belt at 50 AU is not fully understood.

The resonant KBOs are those whose orbital period and that of Neptune form a ratio of integers (whole numbers). The semimajor axis for the 2:1 resonance lies at 47.8 AU, that for the 3:2 resonance at 39.4 AU, and the 4:3 resonance at 36.4 AU. The 3:2 resonance is the most populated, with about 100 members. The semimajor axis of Pluto is 39.48 AU, so it is one of these 3:2 resonance objects. The dynamical similarity between Pluto and the other smaller 3:2 objects has led the latter to be called Plutinos. Some Plutinos have distances of closest approach to the Sun less than 30 AU (i.e., their orbits cross Neptune itself), as does Pluto itself. They can get driven out of their resonant orbit by Pluto and into close encounters with Neptune. This could be one source of short-period comets in the inner Solar System. It has been estimated that there may be as many as 1500 Plutinos larger than 100 kilometres in size. How did these resonances get populated? Perhaps as a result of a net migration of Saturn, Uranus and Neptune outwards as they interacted with planetesimals during the final stages of planet formation. As Neptune moved outwards it could have captured Pluto, along with the other Plutinos, into the 3:2 resonance.

The final group, the scattered KBOs, are characterized by their large, highly eccentric and highly inclined orbits. While their distances of closest approach to the Sun are usually close to 30 AU, their orbits can range out to 270 AU at their farthest point from the Sun. They were probably scattered by Neptune into their far-flung orbits during the planet-formation era. Other planetesimals may have been scattered right out of the Solar System during this phase.

The enormous scientific interest in the Kuiper Belt arises from its almost certainly being the remnant of the Solar System's protoplanetary disk. This has allowed simulations of how the Solar System formed to be enormously improved. Collisions between planetesimals result either in growth through accretion or in destruction by pulverization, depending on the energy of impact. The current Belt perhaps represents only 1% of the original material at this distance from the Sun. There is an obvious link between the Kuiper Belt and debris disks around other stars. *COBE* measurements have been used to set a limit on the total luminosity of the Kuiper Belt at less than one-millionth that of the Sun, so the Solar System is at the very faint end of the debris disk systems detected by *Spitzer.*

Infrared spectra of KBOs show the presence of water ice and silicates on some, while others are featureless. They are expected to have a layered structure, with the more volatile ices concentrated outside the warm, more solid core. Simulations of their formation in the protoplanetary disk suggest that their inner structure may be that of a loose 'rubble pile', similar to what is now believed to be the case for some asteroids and for the nuclei of some comets.

COMETS AND CENTAURS

At the small end of the size distribution of objects in the outer Solar System are the Jupiter-family comets, whose orbits cross or closely approach Jupiter's orbit and which are therefore strongly influenced by the planet. Comet Bowell, which I discussed in Chapter 7, is an example. The nuclei of Jupiter-family comets have an average dynamic lifetime of only 100,000 years before they suffer some strong interaction with planets, and so they must be replenished from some more stable reservoir. It used to be thought that they were long-period comets from the Oort Cloud captured gravitationally by the gas giant planets, but this was shown to be an inefficient process. Instead, it is more likely that they are escaped KBOs that have been scattered inwards by the planets. If this idea is correct, then we would expect to see objects in transit, making the journey from the Kuiper Belt to the inner Solar System. Such objects have now been identified in the population of *Centaurs.* The name was chosen because they behave half as an asteroid and half as a comet. The first was discovered in 1920, but they were not recognized as a distinct population until 1977. Dozens are now known, with orbits between Neptune and Jupiter. Some show comet-like activity, with outgassing and formation of a cometary 'coma'.

It is widely accepted that interaction between newly formed planets and the disk of gas and planetesimals can lead to migration of the orbits of planets. This is the explanation for the 'hot Jupiter' phenomenon, which was such a puzzle when the first exoplanets were discovered. In the 'Nice' model for the evolution of the giant planets of the Solar System, named for the French city where the proposers were based, all four of the outer, giant planets would have originally been much closer to the Sun, with Saturn's orbital period less than twice that of Jupiter. After the gas left over from the formation of the Solar

System had been driven out, Jupiter and Saturn's orbits expanded as a result of their interaction with a massive disk of planetesimals, and the ratio of their orbital periods increased. When the two planets' periods reached a ratio of 2:1, a resonance ensued and their orbits became eccentric. This abrupt transition temporarily destabilized the giant planets, leading to a short phase of close encounters among Saturn, Uranus and Neptune. As a result of these encounters and the interactions of the giant planets with the planetesimal disk, Uranus and Neptune reached their current heliocentric distances and Jupiter and Saturn evolved to their current orbital eccentricities. During the period of resonance, planetesimals would have rained down on the inner Solar System. This dramatic dynamical episode could explain the Late Heavy Bombardment at about 700 million years after the formation of the Solar System, which is so visible on the surface of the Moon, and also the capture of the Trojan asteroids by Jupiter.[11]

In summary, the Solar System we know today is wonderfully more complex than was realized 50 years ago. Over two million minor planets have been catalogued in the Solar System, and the total population of objects larger than, say, 20 kilometres, including comets and all types of asteroids, is believed to be in the billions. The dynamical interaction of these bodies with the planets is complex and dramatic, and their orbits tell us something of the past violent history of the Solar System.

THE SEARCH FOR EXOPLANETS AND HABITABLE PLANETS

For most of my professional life, it seemed that we would never be able to demonstrate the existence of a planet around another star. This was frustrating because astronomers believed that most single stars would be surrounded by a planetary system. Then, in 1995, small variations in the radial velocity of the nearby star 51 Pegasi, observed by Michel Mayor and Didier Queloz of the University of Geneva, showed that it was being orbited by a Jupiter-mass planet.[12] The surprise was that the planet was extremely close to its parent star. Other similar systems were quickly discovered, and these systems became known as 'hot Jupiters'. The theorists had to work hard to understand how such a massive planet could get so close to its parent star, given that they believed the initial formation of these 'hot Jupiters' must have been much farther out. They realized that there is a strong interaction between a newly formed planet and the remaining protoplanetary disk from which it formed, and this can drag the planet in towards the star.

It was natural that the first exoplanets detected should be hot Jupiters because they cause a much stronger perturbation of their parent star than a Jupiter-mass planet farther out. As the sensitivity of the measurements improved, Jupiter-mass planets began to be found farther out from their star. As of October 2011, over 700 extrasolar planets, or exoplanets, were known, and they cover a wide range of distances and masses. Twenty systems are known to have more than one planet. There are now four known cases of a planet orbiting a binary star.

Most extrasolar planets to date have been found with ground-based telescopes after detecting the small variations in the Doppler velocity of the star induced by its planetary companion. The second widely used method for detecting exoplanets is the *transit* method. If the planet crosses the face of its parent star, there is a small dimming of its light, which can be detected through careful observation. A U.K. consortium has had considerable success with this method using their specialized SuperWASP cameras. An advantage of this method is that it can also give the radius of the planet. The space missions *COROT*, launched by the French Space Agency in December 2006, and *Kepler*, launched by NASA in March 2009, both of which are using the transit method, are already dominating the new discoveries.

In rare cases where a star is observed in the process of passing directly in front of another star and amplifying the light of the background star through gravitational lensing, a planet orbiting the lensing star might also cause a small blip of amplification. This has resulted in the indirect detection of several exoplanets.

Here are a few of the milestones in exoplanet discoveries. In 2004, the system 2M 1207b was the first in which the planet was directly imaged, by the European Southern Observatory's Very Large Telescope (Figure 11.3).[13] This was possible firstly because the planet is orbiting a low-mass, faint brown dwarf star and secondly by observing in the infrared, where the planet is more prominent relative to the star than at optical wavelengths. In the same year, the first 'hot Neptune', Mu Arae c, was discovered, which is 14 times the mass of the Earth.

In 2007, the first spectra of exoplanets were measured, and in the same year a team working in the infrared with the *Spitzer Space Telescope* detected water vapour in the atmosphere of an exoplanet.[14] Also in the same year the organic molecule methane was found in the atmosphere of an exoplanet using NICMOS on the *Hubble Space Telescope*.[15]

In 2008, a team using the *Hubble Space Telescope* succeeded in imaging a planet orbiting the bright star Fomalhaut (see Plate XII)

2MASSWJ1207334-393254

778 mas
55 AU at 70 pc

N

E

Figure 11.3. An exoplanet (blob to lower left) orbiting a brown dwarf star. The system 2MASS 1207a,b discovered at the European Southern Observatory's Very Large Telescope in Chile by a team led by Gael Chauvin. This was the first direct image of an exoplanet, made in infrared light. The planet's mass is about 5 Jupiter masses, which is close to the upper limit for planetary masses.

within the dust debris disk that had been first detected by *IRAS*.[16] In the same year, direct imaging of the star HR 8799 with the Keck and Gemini telescopes yielded three exoplanets with masses between 5 and 13 Jupiter masses, resembling a scaled-up version of the outer Solar System.[17]

Of course, one of the main goals of exoplanet research is to find a planet similar to the Earth, and in 2009 the European Southern Observatory announced the discovery of a fourth planet orbiting the nearby star Gliese 581, which had a mass only 1.9 times that of the Earth. This is the lightest exoplanet detected to date and shows that we will soon detect Earth-mass exoplanets.[18]

What does the future hold for exoplanet research? Direct imaging of Earth-mass exoplanets is one of the key goals of the *James Webb Space Telescope* and of planned giant ground-based telescopes, discussed in the next chapter. A more dedicated exoplanet mission is the proposed *Darwin* mission, a flotilla of four free-flying spacecraft, each

Figure 11.4. The future of planet search missions: a flotilla of telescopes working together as an interferometer. This is the U.S. version, *Terrestial Planet Finder*. The European version, *Darwin*, is similar.

containing a 3-metre-diameter infrared telescope, designed to image and analyze Earth-like planets. The light from the telescopes will be combined together to nullify the light from the parent star and allow the planets to be imaged and analyzed spectroscopically. *Darwin* will search for water and oxygen in the planets' atmospheres, which together could indicate the presence of life on the planet. The concept has been studied by the European Space Agency but is not yet in their programme. A similar NASA mission called *Terrestrial Planet Finder* has also been studied but, again, not yet approved (Figure 11.4).

We have seen that there is a great range in planetary system environments, and this may profoundly affect one of the most interesting questions that astronomers face: the probability of life in other planetary systems. An example of a system where life might find it difficult to evolve as it has on Earth is Tau Ceti, the closest analogue of the Solar System. It contains 20 times the mass of the outer Solar System in cool debris particles, a total of about one Earth mass. This system represents a different evolutionary outcome for a Sun-like star with no Jupiter-like planet but many cometary bodies, and thus a potentially heavy and prolonged history of impacts on any inner terrestrial planets. Because Tau Ceti is ten billion years old, life would have had to deal with massive bombardment over very long times-cales. Furthermore, impactors of a size greater than 10 kilometres, capable of causing major extinction events, could arrive at intervals as frequent as a million years or less, making the possibility of an evolutionary history similar to Earth's unlikely.[19] Clearly, the nature of the debris disk surrounding a planetary system has profound implications for whether life can develop on a planet.

The Solar System's debris disk, made up of the asteroid and Kuiper belts, the Oort Cloud and the zodiacal dust cloud, contains the clues to the evolution of our planetary system. Already there are beginning to be interesting dynamical scenarios, such as the 'Nice' model, which connect together these different ingredients and explain anomalies like the Late Heavy Bombardment of the Earth and Moon, and the Trojan asteroids in orbit around Jupiter. So far the debris disks detected around other stars tend to contain much more material and so are not close analogues of the Solar System. Similarly, the exoplanetary systems around other stars tend to have their Jupiter-mass planets nearer in than in our system, and we have not yet managed to detect Earth-mass exoplanets. This is likely to change very soon, and we can expect a big focus on exoplanet research in the next decade. The giant infrared telescopes of the future, described in my final chapter, have as one of their major scientific focuses the detection and characterization of Earth-mass planets. Soon we may be able to finally understand how the Earth and its life formed, and whether this has happened elsewhere.

12

The Future: Pioneering Space Missions and Giant Ground-Based Telescopes

Finally, we look ahead to the next decade of infrared and sub-millimetre astronomy. This will be an era in which the infrared and submillimetre wavebands continue to have a dominant role, with spectacular space missions and giant, new ground-based facilities. I describe the three most recent infrared and submillimetre missions to be launched, *Herschel*, *Planck*, and *WISE*; the future planned missions, the *James Webb Space Telescope* and *SPICA*; and the future ground-based facilities, the Atacama Millimetre/Submillimetre Array and the very large 30–40-metre ground-based telescopes. We will see that the next decade will be just as exciting as the 25 years since the dramatic days of the *IRAS* mission have been.

HERSCHEL AND PLANCK: PROBING THE COLD UNIVERSE

On 14 May 2009, the European Space Agency (ESA) launched together on top of a single Ariane 5 rocket two major space astronomy missions, the *Herschel Space Observatory* and *Planck*. The dual launch of these complex missions, at ESA's space port at Kourou in French Guiana, was a great moment for European space science. It was an extremely moving moment for those of us who had worked on these missions since their inception. They were finally approved by ESA in 1993, *Herschel* (then known as *FIRST*, for Far InfraRed Space Telescope) as the fourth of ESA's Horizon 2000 'cornerstone' missions providing a multi-instrument observatory working at far-infrared and submillimetre wavelengths, and *Planck* (then known as *Cobras-Samba*) as a 'medium' mission to map the cosmic microwave background radiation.[1] *Cobras* and *Samba* had been submitted as separate proposals, but ESA decided they should be merged into a single mission with two instruments, the Low Frequency Instrument (LFI), led by Reno Mandolesi[2] and the

High Frequency Instrument, led by Jean-Loup Puget.[3] Both missions had been studied for many years before 1993. The concept of a very large submillimetre telescope had been proposed as far back as 1982, in the United States by Tom Phillips (as the Large Deployable Reflector, LDR) and in Europe by Thijs de Graauw (as *FIRST*). *FIRST*, which eventually became *Herschel*, was selected as ESA's fourth cornerstone mission in 1984 and finally went ahead as an ESA-led mission, with strong NASA involvement in the development of the instruments.

The progress of the *Herschel* and *Planck* missions was not entirely smooth. On 4 June 1996, the test launch of ESA's new Ariane 5 rocket failed, with the four spacecraft of the *Cluster* magnetospheric mission on board. This caused a crisis within ESA, and there was a strong feeling that *Cluster* should be rebuilt and relaunched. For financial reasons, there was therefore a real possibility that *Planck* would be delayed or cancelled. I organized a joint letter to ESA from Europe's leading cosmologists in support of *Planck*. For the next year, the future of both *Planck* and *Herschel* remained in doubt. An added complication was that ESA wanted to insert a new mission to Mars, *Mars Express*, into the programme, not least because it felt that such a mission would be attractive to the public and to politicians and might help the agency's funding problems. ESA began to consider the idea of merging the *Herschel* and *Planck* missions to save money. It took a while for the eventual concept of two independent missions sharing a common launcher to emerge. Another pressure on ESA's rather tight budget was the idea of a strong European involvement in the *Next Generation Space Telescope (NGST)*, NASA's planned successor to the *Hubble Space Telescope*. Some, like Riccardo Giaconni, who had just taken over as director of the European Southern Observatory, felt that ESA should ditch *Herschel* and go for a much larger stake in the *NGST*. It became clear that the support of the scientific community for *Herschel* needed to be redemonstrated. Goran Pilbratt, the project scientist for *Herschel*, organized an international conference in Grenoble, and I chaired the Scientific Organising Committee. Reinhard Genzel, who had led the group of scientists preparing the science objectives for the *FIRST* proposal in 1993, was a key supporter of the mission at this critical point. The conference, held in April 1997, was a success, and I was able to report a strong level of community support to ESA's Space Science Advisory Committee. ESA firmly committed itself to *Herschel* and *Planck* and remained committed through some difficult times and delays.

Herschel is a pioneering infrared space telescope working in the last unobserved wavelength band. It is the first space observatory to

Figure 12.1. The *Herschel* space observatory.

observe the universe across the full far-infrared and submillimetre range from 50 to 600 microns, wavelengths 100–1000 times longer than that of visible light. *Herschel* penetrates the clouds of dust that shroud newly forming stars and galaxies. The *Herschel* mirror is 3.5 metres in diameter, making it the largest optical or infrared space telescope ever launched (Figure 12.1, Plate XXIII).

Planck is a 1.9×1.5-metre telescope mapping the cosmic microwave background radiation left over from the hot phase of the Big Bang in unprecedented detail, determining cosmological parameters with new

Figure 12.2. The *Planck* surveyor, which is mapping the cosmic microwave background radiation.

precision (Figure 12.2). In a nice complementarity with *Herschel*, *Planck* will also provide the first all-sky survey of the submillimetre sky for point sources, albeit with modest sensitivity. The two missions were launched sitting in the nose cone of the Ariane 5 rocket one above the other, with *Herschel* sitting on top, and travelled together to their remote orbit one million miles from Earth (Plate XXIV). There is a strong overlap in wavelength between the two missions, and both are studying the cold universe, with *Planck* mainly studying the cold relic of the hot Big Bang and *Herschel* the process by which stars and galaxies form. As we have seen throughout this book, stars and planets form in dense clouds of dust and gas and at first no visible light emerges. Only at infrared wavelengths can astronomers see these processes in action. The molecules present in the clouds act as probes of the physical and chemical conditions there.

Following in the steps of *IRAS*, *ISO* and *Spitzer*, *Herschel* represents a huge increase in the size of the telescope mirror, from 0.85 metres for

Spitzer to 3.5 metres for *Herschel*, and in the detail seen at far-infrared wavelengths in cool celestial objects. Additionally, *Herschel* bridges the gap between the wavelengths seen by previous infrared satellites and those studied by submillimetre telescopes on the ground such as JCMT and CSO (Chapter 9). When in the 1990s the *Hubble Space Telescope* made its deep surveys of the distant universe, the Hubble Deep Fields, it saw an entirely new population of distant, irregularly shaped galaxies (see Plate XIV). As we saw in Chapter 9, the James Clerk Maxwell Telescope also looked at these regions at submillimetre wavelengths (850 microns). It, too, saw distant galaxies, but different ones from *Hubble*. *Herschel* operates at wavelengths that bridge the gap between these two instruments and is showing us the relationship between these apparently different galaxy populations.

There had been a number of precursor missions and experiments at submillimetre wavelengths. In the 1990s, French astronomers had an ambitious submillimetre balloon project, PRONAOS, with a 2-metre telescope. NASA launched the SWAS mission with a 70-centimetre telescope studying submillimetre emission lines of water vapour in 1998. There was a similar Swedish-led submillimetre mission launched in 2001 with Canadian, French and Finnish participation, Odin. Archeops was a French CMB experiment flying the HFI instrument from a balloon in 2002. And BLAST, led by the Canadians, flew the SPIRE instrument on a balloon in 2005.

THE SCIENCE OF THE *HERSCHEL* MISSION

Herschel is an enormously versatile space telescope.[4] It is looking at almost all types of cool celestial objects, from our own neighbourhood to the edge of the universe. The nearest objects it is studying are within the Solar System, such as comets, asteroids and Kuiper Belt objects. Further into space, *Herschel* is looking into the dense cocoons of matter that enclose stars in the earliest stages of formation. ESA's *ISO* mission unveiled more than a dozen of these cocoons. *Herschel* is not only finding many more of them but will also have the ability to look inside the giant clouds in space to see the star-forming process happening. *Herschel* is also looking at the rings of debris that accumulate around forming stars. In these dark rings, astronomers believe that planets are completing the process of formation.

One of the prime goals of *Herschel* is to look at young galaxies in the distant universe. Today, galaxies are giant collections of hundreds of billions of stars. The first objects to form in the early universe were much smaller and then grew by merging together in dramatic collisions. These collisions triggered enormous bouts of star formation that *Herschel* is able to see. An unexpected major discovery by *IRAS* was that some galaxies emit the majority of their radiation at infrared wavelengths, almost certainly because we are witnessing major galaxy mergers, but *IRAS* was able to detect only part of this radiation. *Herschel* is making the first census of star-forming galaxies throughout the universe at the epoch of peak star formation, allowing astronomers to chart the star-formation history and evolution of galaxies in the universe. It is revealing the youngest stars in the Galaxy for the first time. It is the first telescope to see, in their entirety, the vast reservoirs of gas that constitute much of the normal matter in the Galaxy.

INTO ORBIT AROUND THE SECOND SUN-EARTH LAGRANGIAN POINT

If *Herschel* were placed in orbit around the Earth, heat from the Earth and Moon would interfere with its instruments, reducing their sensitivity. Instead, *Herschel* orbits a point in space 1.5 million kilometres from the Earth. Called the 2nd Lagrangian point of the Sun-Earth system (or L2 for short; see Chapter 8), it is a local gravitationally neutral point, providing an excellent place for *Herschel* to shelter from the heat being emitted by the Earth, with a good view of the sky. A sun shield protects the telescope from the Sun's radiation, which *Herschel* needs to be bathed in to provide power through its solar arrays.

Three years of routine science operations of *Herschel* are planned. After launch, it took *Herschel* and *Planck* approximately three months to cruise to their operating orbits. During this cruise phase, the instruments were switched on and thoroughly checked, and trials were done to find out how best to make observations. The factor that limits *Herschel*'s lifetime is that its instruments need to be cooled down to almost absolute zero in order to work. This is achieved using 2400 litres of the coolant, liquid helium. As the spacecraft operates, however, the liquid helium is gradually boiling away. When the helium runs out, probably around the end of 2012, the mission will come to an end, as the instruments will warm up and will no longer be able to detect infrared radiation.

THE *HERSCHEL* SPACECRAFT AND INSTRUMENTS

The *Herschel* spacecraft was built all over Europe, with components being contributed by all of ESA's 17 member states. Large academic and industrial consortia from across the world have designed and manufactured the three instruments. They also include contributions from outside the ESA member states by individuals and organizations in the United States, Canada, Poland, China and Taiwan. Over one thousand scientists have been involved in building *Herschel* and its instruments.

Herschel's three instruments are designed to be complementary to one another. The long-wavelength camera SPIRE, led by Matt Griffin, has three bands, centred at 250, 350 and 500 micrometres.[5] SPIRE also contains a spectrometer working at 194–671 microns. The camera takes images whilst the spectrometer splits the infrared light into its equivalent colours. The images show the distribution of matter in the objects observed, such as distant galaxies or individual star-forming clouds in our own Galaxy (Plate XXV). The spectra can be analyzed to show us the chemical compositions and the physical conditions of the objects. There is a small but useful overlap with the PACS camera, which covers the shorter infrared wavelengths from 55 to 210 micrometres. The data from the two instruments can then be used together to cover the entire waveband from 55 to 670 microns – a region of the spectrum that has been only very sketchily studied so far. This wavelength range is essential to see the process of star formation from its beginning, when a cloud of gas begins to collapse together, to its end, when a fully formed star emerges, and to study the outflows from dying stars (Figure 12.3).

The instruments have driven the design of the *Herschel* spacecraft. All of them need an extremely cold environment to work in. *Herschel* therefore has a large cold tank, the cryostat, containing the instruments and around which everything else is placed. At 3.5 metres, the telescope is too large to sit in the cryostat tank, so the telescope cannot be cooled to a point where it emits no infrared radiation itself. Instead, it has been placed behind a large sunshade, like a hi-tech parasol, to shield it from the Sun. This allows the telescope to cool to a tolerable 80 K (−193° C).

PACS, led by Albrecht Poglitsch, moves large, "filled array" detector technology into the far-infrared (55–210 microns) region of the spectrum for the first time. PACS can take images of celestial objects resolved into highly constrained wavelength bands in one shot. Before PACS, such information was gathered at best in a strip across a celestial object, whereas PACS will take an entire image. PACS is again two instruments in one.[6] The first is a camera and the second is a spectrometer. The camera has three broad wavebands and contains 2560 miniature bolometers. The information they collect is combined by

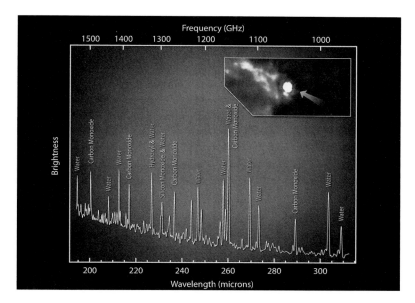

Figure 12.3. Spectrum of the dying red supergiant star VY Canis Majoris made with the *Herschel* SPIRE spectrometer, showing emission lines of carbon monoxide and water. Inset: image made with the SPIRE camera.

computers into colour images. The spectrometer can split the infrared light into its constituents. The spectrometer output can be analyzed to reveal information about how the gas is moving back and forth along the line of sight in the cloud. The spectrometer's information can also reveal the composition of the cloud and its physical characteristics, such as temperature. PACS and SWIRE maps can be combined to give a wonderful impression of the chiaroscuro pattern of dust and illuminating stars in the Milky Way (Figure 12.4 and Plate XXV).

The bolometers at the heart of PACS and SPIRE work by detecting the heat caused when a ray of infrared radiation strikes it. The fainter the source, the colder the bolometers have to be to make an obvious detection. The PACS bolometer and SWIRE detectors must work at the minuscule temperature of 0.3 K. This requires an additional cooling unit, as well as the *Herschel* cryostat that surrounds all the instruments.

The HIFI instrument, led by Thijs de Graauw and Frank Helmich, is essentially a radio receiver designed to tune in to astrophysically important molecules and atoms through the 'fingerprints' they leave at far-infrared wavelengths,[7] at much spectral higher resolution than the spectrometers in SPIRE and PACS. These fingerprints show up either as

Figure 12.4. An early *Herschel* submillimetre image of part of the Milky Way, in the constellation of Vulpecula, made by combining SPIRE and PACS data.

dark lines scattered across an otherwise continuous range of infrared wavelengths or as bright lines against a dark background. The infrared region of the spectrum is full of these lines. Their features, such as their depth and shape, reveal the motions, temperatures and other characteristics of the objects throughout the universe. This in turn will help scientists understand the processes that govern comets, planetary atmospheres, star formation and the development of distant and nearby galaxies. Star formation is thought to be a process that includes the collapse of gas clouds and the expulsion of material as a central star begins to form. By revealing the motion of gas in dense clouds, HIFI is able to pinpoint the locations of star formation and study the complex interplay of collapsing and expanding gas flows.

There are also certain stars that are 'molecular factories'. *Herschel* is investigating them to understand the way they work and what makes them different from other stars. Certain places in space, such as the giant molecular clouds, display tens of thousands of molecular and atomic lines. HIFI has the power to see these lines individually rather than all blurred together. This will give astronomers a unique opportunity to investigate both the chemical composition and the physical state of these locations.

HIFI pushes the technology of submillimetre molecular-line astronomy (Chapter 5) to much shorter wavelengths by using hot electron bolometer detectors at 157–210 microns and SIS receivers at 240–625 microns. These combine the weak signals from space with an artificially generated signal in a detector. The output is a spectrum with a much lower frequency than the original signal. This is known as the heterodyne method and allows the resulting signal to be electronically amplified.

THE *PLANCK* AND *HERSCHEL* SURVEYS

The main goal of *Planck* is to study the cosmic microwave background in unprecedented detail and hence measure cosmological parameters with new precision. One exciting possibility is that *Planck* might detect the signature of dynamical processes in the very early universe, just after the Big Bang. The two *Planck* instruments have a total of nine spectral bands at wavelengths between 350 microns and 1 centimetre. The shortest-wavelength detectors, which use a new fine-mesh 'spider-web' technology, have to be cooled to a temperature of 0.1 K, which is achieved by a three-stage refrigeration process. These detectors are the coolest known objects in space.

Whereas *Herschel* is an observatory that is carrying out many independent observing programmes, *Planck* is a survey mission that scans the whole sky every six months, covering each point on the sky hundreds of times. In July 2011, *Planck* embarked on its fifth complete survey of the sky. In the process, it carried out a survey of extragalactic microwave and submillimetre sources over the whole sky, providing superb maps of the Galaxy at these wavelengths (Figures 12.5 and 12.6, and Plates XXVII and XXVIII). From the point of view of cosmologists studying the cosmic microwave background, these sources and the Galaxy are just foreground noise, which needs to be removed, but to astronomers *Planck* has provided us with the first survey for point sources of the whole submillimetre sky. We are seeing flaring radio sources caused by massive black holes, the cold dust from relatively nearby galaxies and a few very luminous, very distant star-forming

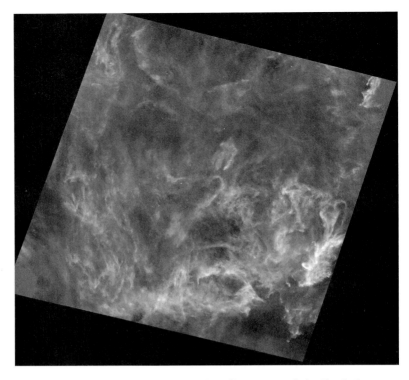

Figure 12.5. Map of interstellar dust filaments made by *Planck* close to the Pole Star, Polaris.

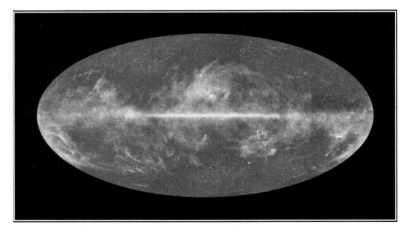

Figure 12.6. *Planck* submillimetre map of the whole sky after the first year of data. The filamentary structure is the emission from the Galaxy, the spots the fluctuations in the cosmic microwave background.

galaxies. Exciting new objects discovered by *Planck* can be followed up with the *Herschel* observatory. *Herschel* is also undertaking some major surveys, but not over the whole sky. Instead selected areas well studied at other wavelengths are being surveyed with SPIRE and PACS. *Herschel* has already generated some fabulous images of parts of the Milky Way as imaged by SPIRE and PACS (Plates XXIX–XXXI).

The two great missions *Herschel* and *Planck* were first approved by the European Space Agency in 1993 and had been studied for many years before that. Their technical complexity has meant that the launch was later than originally planned, but these two missions are a fantastic achievement of European space science with support from other agencies, including NASA. We can expect great insight into the cold and dusty regions from which planets, stars and galaxies form.

WISE

WISE was a small (40 centimetre), cooled infrared telescope launched by NASA on 14 December 2009, led by Ned Wright of the University of California at Los Angeles. Over the following 14 months it carried out two surveys of the whole sky at 3.4, 4.6, 12 and 22 microns to a sensitivity 100 times deeper than *IRAS* at 22 microns and 500 times deeper at 12 microns. The scientific goals were to detect most main belt asteroids larger than 2 kilometres, to find the stars closest to the Sun, which are likely to be brown dwarfs, to detect the most luminous infrared galaxies in the universe, and to provide an essential survey catalogue for the *James Webb Space Telescope*.

During its brief mission, *WISE* discovered 20 comets and more than 33,000 new asteroids, and studied more than 157,000 known asteroids. One of the new asteroids is a Trojan asteroid following the same orbit as the Earth, but luckily always staying at the same distance from us. *WISE* has discovered more than 100 new brown dwarfs so far, six of them Y dwarfs, the coolest category. The coldest of these, WISE 1828+2650, is the coldest known brown dwarf, with a temperature of 298 K (25 °C), close to the temperature of the Earth. The Y dwarfs are at distances between 9 and 40 light years from the Sun, making them some of our closest neighbours in the Milky Way Galaxy. The *WISE* survey also found hundreds of millions of stars and galaxies, which are still being studied and will provide a vital resource for future surveys and space missions.

SPICA

SPICA is a proposed Japanese-led mission with a strong European involvement, planned to be launched in the early 2020s, and consists of a cooled 3.5-metre telescope operating at 4–200 microns. The telescope area is 16–36 times larger than those of *IRAS*, *ISO* and *Spitzer*, providing a huge increase in sensitivity and resolution. The cooling of the mirror means that *SPICA* would outperform *Herschel* at 50–100 microns. The mission would study the formation of planets, stars and galaxies, and would carry out deep cosmological surveys.

THE *JAMES WEBB SPACE TELESCOPE*

The *James Webb Space Telescope* (*JWST*), named after a NASA administrator (it was originally known as the *Next Generation Space Telescope*), is a 6.5-metre telescope due to be launched on an Ariane rocket in 2018. It is a joint project of NASA and the Canadian and European space agencies. It has a sun shield the size of a tennis court and will operate from 0.6 to 27 microns (Figure 12.7). It has four instruments: a near-infrared (0.6–5 microns) camera, a near-infrared multi-object spectrograph, a mid-infrared (5–27 microns) imager and spectrograph, and a fine-guidance sensor that images over wavelengths from 1.6 to 4.9 microns to acquire guide stars and control the *JWST* pointing. It is intended as the successor to the *Hubble Space Telescope* but should perhaps be thought of as the successor to *ISO* and *Spitzer*, given its emphasis on infrared instrumentation. The shift towards the infrared is essential to meet the main science themes, which are to detect the first bright objects that formed in the universe (the end of the 'dark ages') and to study the assembly and evolution of galaxies, the birth of stars and protoplanetary systems, and planetary systems and the origins of life.

ATACAMA LARGE MILLIMETRE ARRAY (ALMA)

The Atacama Large Millimetre/Submillimetre Array (ALMA) is the most ambitious ground-based telescope constructed to date and is due for completion at an altitude of 5 kilometres in the Chilean Andes in 2013. ALMA is a global endeavour – a partnership of Europe, North America and East Asia in cooperation with the Republic of Chile. It consists of an array of 66 reconfigurable antennas with a data processor capable of ten million billion (10^{16}) operations per second and receivers covering

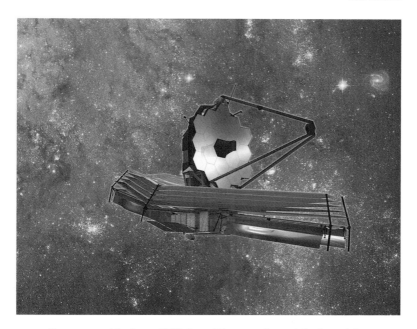

Figure 12.7. The *James Webb Space Telescope*, planned for launch in 2014.

every atmospheric window between 0.3 and 9.6 millimetres, providing an impressive combination of sensitivity, angular resolution, spectral resolution and imaging fidelity (Figure 12.8). By moving the antennas, the effective diameter of the array can be varied between 150 metres and 16 kilometres. Compared with the James Clerk Maxwell Telescope it will have a much smaller field of view, but it will have incomparably better sensitivity and spatial resolution.

In the nearby universe, ALMA will study the processes of star and planet formation, revealing the details of young, still-forming stars, and is expected to show young planets still in the process of developing from dust debris disks. ALMA will also explore the complex chemistry of the giant clouds of gas and dust that spawn stars and planetary systems. On the cosmological scale, ALMA will study the first stars and galaxies that emerged from the cosmic dark ages.

The first science results were announced in October 2011 based on the first 12 antennas and included a spectacular millimeter-wavelength (1–3 millimetre) image of the 'Antennae' interacting galaxies (Plate XIII). ALMA was preceded by APEX, the Atacama Pathfinder Experiment, a 12-metre submillimetre telescope operated from 2003 at the ALMA site by German and Swedish astronomers in collaboration with ESO.

Figure 12.8. The Atacama Large Millimetre/Submillimetre Telescope.

EXTREMELY LARGE TELESCOPES: TMT, GMT AND E-ELT

Three giant optical and infrared ground-based telescopes with mirror diameters in the range 30–40 metres are planned for the end of the decade, two by the United States and one by Europeans.

The largest is the European Extremely Large Telescope (E-ELT), planned for 2018 as a ground-based segmented-mirror telescope with a startling overall diameter of 39 metres (Figure 12.9). Its focus is on the wavelength range 0.36–24 microns, similar to that of the *James Webb Space Telescope*, but its huge light-gathering power means that it will go far beyond the performance of the *JWST*, at least within the wavelength windows accessible from the ground. The scientific case for the E-ELT is similar to that for the *JWST*, to study the very first stars and galaxies in the universe at redshifts 10 and beyond, to study the evolution of black holes over the whole of cosmic time, and to image and take spectra of Earth-mass exoplanets around nearby stars. It will use adaptive optics to overcome the twinkling effect of the Earth's atmosphere. A deformable secondary mirror is used to compensate for the distortions in the Earth's atmosphere, which are monitored by observing a bright star or by shining a laser in the direction of observation.

In the United States, two similar large telescopes are planned, GMT and TMT, on the same timescale. The Thirty Meter Telescope (TMT)

Figure 12.9.   The proposed European Extremely Large Telescope, with a mirror diameter of 39 metres.

is a collaboration between the California Institute of Technology, the University of California, and Canadian, Chinese, Indian and Japanese astronomers. It will be built on Mauna Kea, Hawaii, and is due for completion in 2018. Its first-light instruments would work at optical and near-infrared wavelengths.

The Giant Magellan Telescope (GMT) will consist of seven 8.4-metre-diameter telescopes operating together in a cluster and thus equivalent to a single 24.5-metre telescope. Also planned for 2018, it is being built by a consortium of American, Australian and Korean institutions and will be sited at Las Campanas, Chile. Its four instruments will between them cover the wavelength range 0.4–28 microns. The combination of the *JWST* and these three giant ground-based telescopes promises a remarkable period for optical and infrared astronomy.

Other future facilities planned in the infrared and submillimetre wavebands include the Cerro Chajnantor Caltech Atacama Telescope (CCAT), a 25-metre submillimetre single-dish telescope to go to Chile in 2017. The South Pole Telescope, a 10-metre telescope in Antarctica designed for cosmic microwave background studies, began work in 2007 and also has great potential for submillimetre astronomy. A powerful new millimetre array, CARMA, made by combining dishes from the Owens Valley Radio Observatory (OVRO) and Hat Creek millimetre arrays, started work at a site near OVRO in 2006. *SOFIA* (the *Stratospheric Observatory for Infrared Astronomy*), a NASA-German airborne observatory carrying a 2.5-metre telescope and operating across the whole infrared and submillimetre range from 0.7 microns to 1.6 millimetres, made its first science flight in November 2010, carrying out spectroscopy in the Orion Nebula and other star-forming regions.

Thus the infrared and millimetre wavebands will be very much the focus of the ground-based astronomy of 2020, with the thrust towards the very highest redshifts at one extreme and towards studying terrestrial-mass planets at the other. Of the five largest ground-based telescopes planned, only one will not be working at infrared and submillimetre wavelengths. This is the Square Kilometre Array (SKA) radio telescope, a huge international project that will revolutionize radio astronomy. The other giant telescopes, ALMA, TMT, GMT and E-ELT, demonstrate how far the field of infrared astronomy has developed since its invention by William Herschel, its very slow development in the nineteenth century, and the pioneering work of the 1960s, 1970s and 1980s. Our night vision has expanded unbelievably in its grasp and power, and the vision of the universe it reveals is still unfolding.

Epilogue

I find the slow emergence of infrared astronomy a moving story. It began with the revolutionary discovery by the self-taught William Herschel in 1800 of invisible radiation from the Sun, the significance of which took so long to be fully appreciated. Reading his methodical and imaginative papers makes his genius clear. Piazzi Smyth made the next step, with detection of infrared radiation from the Moon in 1856. I found his book about his expedition to Tenerife captivating, one of the great works of scientific popularization from the Romantic era. There followed another 50 years of painstaking work trying to detect the brightest stars and planets in infrared light and the slow progress of stellar infrared astronomy in the first half of the twentieth century, culminating in the work of Harold Johnson and his group. In 1930, Robert Trumpler discovered the key ingredient for understanding the infrared sky: interstellar dust.

We then come to the titans of the modern era, Frank Low and Gerry Neugebauer, repeatedly being told they were wasting their time as they tried to push astronomy into the infrared. They and their colleagues from many different groups deserve immense credit for their pioneering work in the 1960s and 1970s. This led to the explosive development of infrared astronomy following the launch of the *IRAS* satellite in 1983. It was such an exciting time to see the clouds of interstellar dust directly shining at us in infrared light and to find some of the most distant galaxies known at that time, infrared monsters convulsed in huge bursts of star formation. Considering where infrared astronomy had been only a decade or so earlier, it was wonderful to be using *IRAS* as a cosmological probe, linking the infrared galaxy distribution to the cosmic microwave background radiation left over from the Big Bang itself. *IRAS* was followed by ever more sophisticated space missions: *ISO*, *Spitzer*, *Akari* and today *Herschel*. And then there have been the giant ground-based telescopes, working either in the optical and near infrared or at submillimetre wavelengths. To be searching in the 1990s for submillimetre sources one thousand times fainter than those we had been trying to observe only 20 years earlier seemed like magic.

Our own solar system is an infrared world, permeated with dust and with dusty debris such as comets and asteroids. And this links us to debris disks around other stars and to exoplanets. This is perhaps the most exciting area of astronomy today, and it cannot be long before the giant space and ground-based telescopes being built or planned discover and characterize planets like the Earth. There is a widespread expectation amongst the public and many scientists that planets with life, and indeed intelligent civilizations, are common in the Galaxy. It will be startling if we find evidence for this. It will be equally shocking to become convinced that we are alone. That is another story.

I was lucky to be involved in many of the modern developments and discoveries of infrared astronomy and to have met almost all of the leading players. It is only through writing this book that I have become fully aware of the amazing time we have lived through.

Notes

PREFACE

1 Allen 1975.
2 Low, Rieke and Gehrz 2007.
3 Rieke 2009.
4 Longair 2006.

1. INTRODUCTION

1 Conway Morris 2003.
2 Maxwell 1864.

2. WILLIAM HERSCHEL OPENS UP THE INVISIBLE UNIVERSE

1 Armitage 1953; Hoskins 2011.
2 Astronomers capitalize "Galaxy" when referring to our own.
3 Hoskins 1963. In his two wonderful books on William Herschel, Michael Hoskins seems to me to seriously underestimate the significance of Herschel's discovery of infrared radiation.
4 Holmes 2008.
5 W. Herschel 1800a.
6 W. Herschel 1800b.
7 W. Herschel 1800c.
8 Herschel's experiments on radiant heat were continued by the Italian scientist Macedonia Melloni (1798–1854). Melloni also reported in an 1846 letter that in 1830 he had observed the Moon through a large lens and detected heat from it using a 'thermomultiplier' (Price 2009).
9 Holmes 2008.
10 The French physicist Claude-Servais Pouillet carried out a very similar experiment at about the same time (Pouillet 1838).
11 J. F. W. Herschel 1840.
12 This result was confirmed in 1847 by Armand Fizeau (1819–1896) and Léon Foucault (1819–1868). They also showed that infrared rays show the phenomena of interference, polarization and diffraction, which were already known for visible light (Fizeau and Foucault 1847).
13 Holmes 2008.

3. 1800–1950: SLOW PROGRESS – THE MOON, PLANETS, BRIGHT
STARS AND THE DISCOVERY OF INTERSTELLAR DUST

1 Seebeck 1826.
2 The Royal Observatory Edinburgh moved to its present site on Blackford
   Hill in 1896. In 1853, Piazzi Smyth was responsible for installing the 'time
   ball' on top of Nelson's Monument in Edinburgh to give a time signal to
   ships at Edinburgh's port of Leith, and in 1861 this visual signal was aug-
   mented by the 'One-O'clock Gun' at Edinburgh Castle, which visitors to
   Edinburgh can still hear today.
3 The Canary Islands authorities continue to support astronomy to the present day. The Imperial College Infrared Flux Collector was sited on Tenerife,
   not far from where Piazza Smyth observed, in the 1970s and is still in use
   today. However, the main thrust of astronomy in the Canary Islands has
   shifted to the Observatory of La Roque dos Muchachos on La Palma, where
   there are a range of telescopes, including the United Kingdom's 4-metre
   William Herschel Telescope and the new Spanish Gran TeCan 10-metre telescope. Exterior lighting throughout the island of La Palma is carefully con-
   trolled to ensure the darkest possible site at the observatory.
4 Piazzi Smyth 1858.
5 Parsons 1873.
6 Huggins 1868.
7 Another early claim of detected infrared radiation from stars was by E.F.
   Stone in 1879. He used a pair of matched thermocouples at the telescope
   focus to compensate for thermal drifts in the instrument and fluctuations
   in the emission from the sky, and claimed detections of Arcturus and Vega
   (Stone 1870). In 1878, the famous American inventor Thomas Edison used an
   instrument he called a *tasimeter*, which was essentially a bolometer consisting of a pellet of finely divided carbon powder, to study a total solar eclipse
   in the infrared. Edison also claimed to have detected the star Arcturus (Eddy
   1972). However, the claims of Huggins and Edison to have detected infrared radiation from stars were disputed by C.V. Boys (Boys 1890). Boys built
   a new instrument, which he called a *radiometer*, consisting of two matched
   thermocouples suspended in a current loop inside a magnet. A small mirror attached to this torsional pendulum allowed accurate measurement of
   deflections using a reflected light beam. Boys demonstrated that his detector was more sensitive than those of Huggins and Edison, but he could not
   detect infrared radiation from bright stars.
8 Langley 1886.
9 Langley 1900.
10 Nichols 1901. The Nichols radiometer is sometimes confused with the
   better-known Crookes radiometer, a novelty device that operates under the
   action of molecular pressure in a partially evacuated bulb.
11 Abbott 1924.
12 Coblentz 1914, 1922.
13 Menzel, Coblentz and Lampland 1926.
14 Rieke 2009.
15 Pettit and Nicholson 1928, 1933.
16 Pettit and Nicholson 1936.
17 Pettit and Nicholson 1930, 1935, 1940.
18 Wesselink 1948.
19 Kuiper, Wilson and Cashman 1947.
20 Rieke 2009.

21 Fellgett 1951.
22 Trumpler 1930.
23 Stebbins, Huffer and Whitford 1939; Whitford 1958.
24 Greenberg 1963. This article appeared in the first issue of what was to become the major astronomy review journal, *Annual Review of Astronomy and Astrophysics*.
25 Lindblad 1935. This idea was developed further by Oort and van der Hulst (1946).
26 Platt 1956.
27 Cayrel and Schatzman 1954; Hoyle and Wickramasinghe 1962.
28 Eddington 1925.
29 Helium, lithium, beryllium and boron are known as the light elements.
30 Burbidge, Burbidge, Fowler and Hoyle 1957; Cameron 1957.

4. DYING STARS SHROUDED IN DUST AND STARS BEING BORN:
THE EMERGENCE OF INFRARED ASTRONOMY IN THE 1960S AND
1970S

1 Johnson 1962. Eric Becklin proposed also the H band at 1.65 microns.
2 Other early pioneers in near-infrared astronomy using lead sulphide cells were the French astronomer M. Lunel, who measured 61 stars in H, K and a band which combined I and J (Lunel 1960), and the Russian Vasili Moroz (1931–2004), who drew on the inspiration of radio astronomy to study the Crab Nebula, the Galactic centre, and Orion (Moroz 1960, 1961). However, Moroz's primary interest was the study of the planets, and by 1970 he was concentrating on the activities of the Soviet space programme.
3 Low 1961a, b. Liquid helium (helium-4) at atmospheric pressure boils at 4.2 K. In fact, to operate the Low bolometer, a temperature of 2 K (−271 °C) is needed and so the helium must be pumped to low pressure. Helium-3 (an isotope of helium with one fewer neutron in its nucleus) boils at 3.2 K; it is the coldest known liquid at atmospheric pressure. Helium-3 can be pumped to a temperature of 0.25 K, and the dilution refrigerator on the *Planck* satellite (Chapter 12) achieves a temperature of 0.1 K with a mixture of helium-4 and helium-3.
4 Low and Johnson 1964.
5 Low, Rieke and Gehrz 2007.
6 Low and Johnson 1964.
7 Johnson 1966a.
8 Johnson 1965.
9 Shu-Shu Huang, in 1969, was the first of a series of theoreticians to try to model the emission from these dust shells (Huang 1969a, b).
10 The first attempt at a near- and mid-infrared survey had been made by Freeman Hall in 1964 (Hall 1964), and Neugebauer and Leighton acknowledged this survey as the inspiration for their own survey. The telescope used for the Two Micron Survey was later relocated to White Mountain, California, and used for submillimetre observations. It is now in the Smithsonian Museum.
11 Neugebauer and Leighton 1969. The Two Micron Catalogue contained 5612 sources between declinations −33° and +81° with K magnitude brighter than 3.0.
12 Neugebauer, Martz and Leighton 1965.
13 The following year Neugebauer, Leighton and their co-workers gave details of a further 14 interesting sources from the survey. These became known

as CIT1–14, where CIT stands for California Institute of Technology. CIT1 and CIT10 were invisible at optical wavelengths even with a large telescope. All 14 objects were stars with circumstellar dust shells (Ulrich et al. 1966).

14 Most objects in the survey (93%) could be identified in the 1943 Dearborn Catalog of faint red stars (Grasdalen and Gaustad 1971).

15 Neugebauer, Becklin and Hyland 1971.

16 This second phase of ascent of the giant branch is called the asymptotic giant branch.

17 On the theory side, Wayne Stein wrote some important early papers on far-infrared emission from interstellar grains and on models of dusty H ɪɪ regions (Stein 1966a, b, 1967).

18 Gillett, Low and Stein 1967, 1968a, b.

19 R. D. Gehrz, 'The History of Infrared Astronomy: the Minnesota-UCSD-Wyoming Axis'.

20 Woolf and Ney 1969.

21 Gilman 1969.

22 In 1973, Ed Ney and his colleagues published a paper entitled 'Cygnids and Taurids, Two Classes of Infrared Objects' (Strecker, Ney and Murdock 1973). The 'Cygnids' were those that like NML Cyg showed the 10-micron infrared feature, while the 'Taurids' were those like NML Tau that did not. The latter were candidates to be stars with dust shells containing graphite grains. Ironically, the reason NML Tau does not show the silicate excess turns out to be that its dust shell is particularly thick. It is in fact oxygen-rich. The prototype carbon star with a thick circumstellar dust shell turned out to be not NML Tau but another Two Micron Survey source, IRC +10216.

23 Low and Krishna Swamy 1970.

24 Gehrz and Woolf 1971.

25 In 1966, the Mexican astronomer Eugenio Mendoza found that T Tauri stars are strong infrared sources and in fact are emitting much of their energy in the infrared (Mendoza 1966, 1968).

26 Becklin and Neugebauer 1967.

27 Soon afterwards, Doug Kleinmann and Frank Low found extended 20-micron radiation around the BN object (Kleinmann and Low 1967), and Ney and Allen found another extended infrared object nearby (Ney and Allen 1969). Finally, in 1970 Low and Aumann mapped the whole region around the Orion Nebula at 50–300 microns and estimated a total infrared luminosity of 200,000 solar luminosities (Low and Aumann 1970). Two other H ɪɪ regions to be mapped in the mid-infrared at this time were M8 (Gillett, Low and Stein 1968) and M17 (Kleinmann 1970).

28 Herbig 1962.

29 Mezger and Smith 1977.

30 Precession of the equinoxes is caused by the Moon's gravity pulling on the Earth's equatorial bulge. As a result, the Earth's North Pole precesses in a large circle around the sky once every 26,000 years. Astronomers need to correct for this effect when pointing their telescopes at the stars.

31 Quoted in Low, Rieke and Gehrz 2007.

32 Becklin and Neugebauer 1968.

33 A similar infrared source was found in the centre of our nearby companion galaxy Messier 31, the Andromeda Galaxy, by Sandage, Becklin and Neugebauer in 1969 (Sandage, Becklin and Neugebauer 1969).

34 Johnson had published 0.4–3.5-micron observations of the central region of 10 bright galaxies in 1966, made with much larger telescopes than that

used for the Two Micron Survey, and found the emission could be modelled in terms of starlight (Johnson 1966).

35 Low and Johnson 1965.

36 Seyfert 1943.

37 Kleinmann and Low 1970a; Oke, Neugebauer and Becklin 1970. Rees et al. (1969) proposed, correctly, that the infrared emission from Seyferts could be caused by dust illuminated by a central ultraviolet source of radiation.

38 Rieke and Low 1972.

39 Rieke and Low 1975.

40 Gillett, Forrest and Merrill 1973.

41 Gillett, Low and Stein 1967, 1968a, b.

42 Walker and Price 1975.

43 Kleinmann, Gillette and Joyce 1981; Allen et al. 1977.

44 Price and Walker issued a new, improved version of their catalogue in 1976 (Price and Walker 1976), which became known as the AFGL Catalogue (AFGL stood for Air Force Geophysical Laboratory, which was the new name of their laboratory). Ground-based follow-up studies showed that the reanalysis had been largely successful in removing the spurious sources. A detailed account of early infrared surveys of the sky, made from the ground and from satellites by the U.S. Air Force, is given by Steve Price (Price 2009). Rocket flights were also used by the Cornell and NRL groups to make mid-infrared observations in the late 1960s and early 1970s (see Chapter 5). There were three further Air Force rocket infrared surveys prior to the launch of *IRAS*: *SPICE* (1979), *FIRSSE* (1982) – a precursor to *IRAS* studying the wavelength range 10–100 microns – and *ZIP* (1980), studying the infrared emission from zodiacal dust (Murdock and Price 1985).

45 Yorke 1977.

46 Leung 1975, 1976.

47 They did, however, make some rather strong simplifying assumptions to speed up their computer codes, and I began to wonder if these assumptions were valid. I developed my own code, which while rather slow and cumbersome did treat the flow of radiation accurately (Rowan-Robinson 1980).

48 Rowan-Robinson and Harris 1983.

49 Knapp 1985.

50 Greenberg 1963; Lynds and Wickramasinghe 1968; Aanestad and Purcell 1973.

51 Savage and Mathis 1979.

52 Mathis, Rumpl and Nordsieck 1977.

53 Draine and Lee 1984; Laor and Draine 1993; Weingartner and Draine 2001; Draine 2003.

54 In his 2003 review, Bruce Draine argues that because the timescale for destruction of dust grains in shocks from supernovae is short, most grains cannot originate from circumstellar dust shells but must be formed in interstellar dust clouds. While it is generally accepted that grains grow molecular and ice mantles in interstellar clouds, it is harder to see how silicate and graphite grain cores could condense there. Anders and Zinner (1997) showed that primitive meteorites contain pristine interstellar grains, many of which originated in red giant stars.

55 Sellgren 1984.

56 Leger and Puget 1984; Puget, Leger and Boulanger 1985; Allamandola, Tielens and Barker 1985; Puget and Leger 1989. Duley and Williams (1981) had been the first to discuss a possible role for aromatic hydrocarbons in explaining the unidentified infrared bands, though they later focused on an

alternative explanation, amorphous carbon in which additional hydrogen atoms had been inserted.

57 Allamandola, Tielens and Barker 1989; Allamandola and Hudgins 2003.
58 Jennings 1986.
59 Storey 2000.
60 Joseph, Meikle, Robertson and Wright 1984; Joseph and Wright 1985; Wright, Joseph and Meikle 1984.
61 Joseph, Wade and Wright 1984.

5. BIRTH OF SUBMILLIMETRE ASTRONOMY: CLOUDS OF DUST AND MOLECULES IN OUR GALAXY

1 Low 1961a, b.
2 Low, Rieke and Gehrz 2007.
3 Aumann, Gillespie and Low 1969.
4 Low and Tucker 1968.
5 Low and Aumann 1970.
6 A more eccentric Frank Low paper discussed 'irtrons' powered by continuous creation of matter in the centres of galaxies, a response to the surprisingly high infrared luminosities of some galaxies (Low 1970). We now know this is caused by bursts of massive star formation.
7 Bastin et al. 1964. Earlier 1–3 millimetre observations of the Sun and Moon had been made by William Sinton in 1953–6 in Baltimore.
8 Beckman, Bastin and Clegg 1969.
9 Park, Vickers and Clegg 1970.
10 Gaitskell, Newstead and Bastin 1969. Their prescient abstract is worth quoting: 'The detection of solar radiation in the 860 micron window at a sea-level site (London, England) is reported.… The possibility of submillimetre astronomy from low altitude sites is discussed. We conclude that most of the world's observatories in temperate or cool climates could make worthwhile measurements in the 730 and 860 micron windows for a limited period each year. Measurements at 350 and 450 microns should be possible from a more limited number of sites in cold climates. Low temperature seems to be the most important criterion in choosing a site with low precipitable water-vapour content.' In fact, 850-micron measurements today are only attempted from high-altitude observatories, because such sites can be used for submillimetre measurements all year round.
11 Clegg, Ade and Rowan-Robinson 1974. I wrote in *New Scientist* in 1974: 'One of the last large astronomical gaps in the electromagnetic spectrum is the submillimetre decade of wavelengths from 100 micrometres to 1 millimetre. It is particularly important because for many of the most interesting objects in the universe (quasars, 'active' galaxies, and even the nucleus of our own Galaxy) this is the band where the bulk of their radiation is emitted. … Over the past couple of years, my colleagues, Peter Clegg and Peter Ade, and I have been making observations in this region and we have increased the number of extragalactic objects detected at submillimetre wavelengths from 3 to 8, so opening up the prospect of important contributions to astrophysics and cosmology from these wavelengths.'
12 Quoted in Low, Rieke and Gehrz 2007.
13 Pipher 1971; Soifer, Houck and Harwit 1971; Houck, Soifer, Pipher and Harwit 1971.
14 Gould and Harwit 1963.

15 Davidson and Harwit 1967.

16 Hoffmann and Frederick 1969.

17 Hoffman, Frederick and Emery 1971.

18 Jennings and Moorwood 1971; Furniss, Jennings and Moorwood 1972a, b, 1975; Emerson, Jennings and Moorwood 1973. A Dutch group also flew far-infrared balloon experiments in 1973 (Olthof and van Duinen 1973).

19 Joyce, Gezari, and Simon 1972; Gezari, Joyce and Simon 1973.

20 Fazio et al. 1974.

21 Telesco, Harper and Loewenstein 1976.

22 Werner et al. 1975; Elias et al. 1978. These observations were made using a germanium detector developed by Mike Hauser, which became the proto-type for the detectors on the FIRAS instrument on the *COBE* mission.

23 Rieke et al. 1980.

24 Wilson, Jefferts and Penzias 1970. Like the other famous paper by Penzias and Wilson, on the discovery of the cosmic microwave background, this paper is only just over 1 page long.

25 In 1971, Penzias, Jefferts and Wilson detected two isotopes of CO. An iso-tope of an atom is a form of the atom with more or fewer neutrons in the nucleus of the atom than in the naturally occurring form. Carbon always has six protons in its nucleus, and the normally occurring form, ^{12}C, has six neutrons, but it can also be found with seven neutrons (^{13}C) and eight neutrons (^{14}C), where the superscript number denotes the total number of nucleons (protons plus neutrons). The normal form of CO is $^{12}C^{16}O$, and Penzias and his collaborators had detected $^{13}C^{16}O$ and $^{12}C^{18}O$. The value of these isotopes turned out to be that their line radiation penetrated deeper into the clouds of molecules than normal CO (Penzias, Jefferts and Wilson 1971).

26 In 1971, Penzias, Solomon, Wilson and Jefferts announced the discovery of interstellar carbon monosulfide, CS (Penzias, Solomon, Wilson and Jefferts 1971), and also detected CO, CN and CS in the carbon-rich circumstellar dust shell IRC +10216 (Wilson, Solomon, Penzias and Jefferts 1971). Also in 1971, the same authors detected carbonyl sulfide (OCS), and in 1973 methyl cyanide (CH_3CN), isocyanic acid (HNCO) and cyanoacetylene (HC_3N). L.E. Snyder and D. Buhl detected hydrogen cyanide (HCN) in 1971 (Snyder and Buhl 1971).

27 McKellar 1940.

28 Shklovsky 1952.

29 Townes and Schawlow 1955.

30 Weinreb et al. 1963.

31 Cheung et al. 1968.

32 Snyder, Buhl, Zuckerman and Palmer 1969.

33 Scoville, Solomon and Penzias 1975.

34 These include cyanogen (CN), carbon monosulfide (CS), hydrogen cya-nide (HCN), hydrogen icocyanide (HNC), the ethynyl radical (C_2H), sili-con monosulfide (SiS), silicon monoxide (SiO), methane (CH_4), ammonia (NH_3), cyanoacetylene (HC_3N), cyanodiacetylene (HC_5N), cyanotriacety-lene (HC_7N), butadynyl (C_4H), cyanoethynyl (C_3N) and methyl cyanide (CH_3CN).

35 Phillips and Jefferts 1973.

36 Phillips, Jefferts and Wannier 1973.

37 Phillips and Rowan-Robinson 1978.

38 A later meeting at Queen Mary chaired by Malcolm Longair discussed whether the site for this telescope should be at La Palma or Mauna Kea.

Although I wasn't at this meeting, Longair did ask me where I thought the telescope should go. I said that to put a submillimetre telescope on La Palma would be a complete mismatch of telescope and site; the telescope had to be on Mauna Kea.

39 Dame et al. 1987.
40 Dame, Hartmann and Thaddeus 2001.
41 Blitz 1979.
42 Huggins et al. 1975.
43 Rickard et al. 1975. These galaxies had been detected at radio frequencies in the hydroxyl (OH) molecule by Weliachew (1971).
44 Morris and Rickard 1982.
45 Young and Scoville 1991.

6. THE COSMIC MICROWAVE BACKGROUND, ECHO OF THE BIG BANG

1 For example, Mather and Boslough 1996; Smoot and Davidson 1993; Rowan-Robinson 1993; Chown 1996; Peebles, Page and Partridge 2010.
2 An object at the horizon would be receding at the speed of light. Note that the relation redshift = v/c only works when the recession speed v is much less than the speed of light c.
3 Penzias and Wilson 1965a. Arno Penzias and Bob Wilson shared the Nobel Prize for Physics in 1978 for this momentous discovery.
4 Dicke, Peebles, Roll and Wilkinson 1965.
5 Roll and Wilkinson 1966.
6 Penzias and Wilson 1965b.
7 Smoot, Gorenstein and Muller 1977. The dipole anisotropy had been predicted by Dennis Siama in 1967. In his 1971 textbook *Physical Cosmology*, Jim Peebles emphasized the similarity between the cosmic background radiation and the nineteenth-century concept of the aether. When the Smoot et al. paper was published in 1977, I wrote a report in *Nature* headed 'Aether drift detected at last'.
8 Wagoner, Fowler and Hoyle 1967.
9 Clegg, Newstead and Bastin 1969.
10 Robson et al. 1974. Dirk Muelner and Ray Weiss of MIT had made balloon-borne measurements in 1973 of the CMB spectrum between 1 and 3 millimetres, which were consistent with a 2.7 K blackbody (Muelner and Weiss 1973).
11 Woody, Mather, Nishioka and Richards 1975; Woody and Richards 1979.
12 Matsumoto et al. 1988.
13 Gush, Halpern and Wishnow 1990.

7. THE *INFRARED ASTRONOMICAL SATELLITE* AND THE OPENING UP OF EXTRAGALACTIC INFRARED ASTRONOMY: STARBURSTS AND ACTIVE GALACTIC NUCLEI

1 A history of the development of *IRAS* has been given by G.M. Smith and G.F. Squibb (1984, NASA Document 19840035078). Russ Walker chaired the Joint Mission Definition Team set up by NASA in 1975 to study the idea of the joint U.S.-Dutch mission. The joint *IRAS* project was approved by the

Dutch in December 1976 and by NASA in January 1977. Formally Gerry Neugebauer and Reinder van Duinen (and later Harm Habing) were described as co-chairmen of the Joint *IRAS* Scientific Working Group (JISWG).

2 The *IRAS* satellite and mission are described in a paper led by Gerry Neugebauer and coauthored by the *IRAS* science team (Neugebauer et al. 1984).

3 Aumann et al. 1984.

4 Walker et al. 1984.

5 Davies et al. 1984. A more complete list of *IRAS* asteroids was compiled by Tedesco, Noah, Noah and Price (2002). Mark Sykes and his collaborators discovered dust trails along the orbits of periodic comets resulting from debris shed by the comets (Sykes, Lebofsky, Hunten and Low 1986; Sykes and Walker 1992).

6 An overview of the *IRAS* Minisurvey was given in Rowan-Robinson et al. 1984.

7 Beichman et al. 1984.

8 In March 1983, Tom Soifer and I circulated a memo to the *IRAS* Science Team on 'Evidence that galaxies are major *IRAS* science', based on 70 identified galaxies, and this developed into one of the *IRAS* papers, led by Soifer (Soifer et al. 1984a). Thijs de Jong also led a study of *IRAS* observations of bright optical galaxies, which showed that almost all the *IRAS*-detected galaxies were spiral galaxies (de Jong et al. 1984).

9 Houck et al. 1984.

10 Soifer et al. 1984b.

11 Miley et al. 1984.

12 Later Miley and his Dutch colleagues studied *IRAS* emission from Seyfert galaxies (de Grijp, Miley, Lub and de Jong 1985), and he, Gerry Neugebauer and others studied *IRAS* emisison from quasars (Neugebauer, Miley, Soifer and Clegg 1986).

13 Helou, Soifer and Rowan-Robinson 1985; a similar result was found by Thyjs de Jong and his collaborators (de Jong, Klein, Wielebinski and Wunderlich 1985).

14 Low et al. 1984.

15 In the paper on the Galactic centre led by Nick Gautier that we wrote during the summer of 1984 (Gautier et al. 1984), we said: 'The bright source at l=357.31, $b = -1.34$ in the 12 and 25 micron images is unidentified. It has flux densities of … in the 12, 25, 60 and 100 micron bands respectively. These correspond to a color temperature of 220 K between 12 and 60 microns. … There is no indication in the *IRAS* data that this source is extended. … The fixed position of the source over a period of a week makes it unlikely to be within the solar system. The object had not been seen in other infrared surveys because they either did not cover this part of the sky or were not sensitive enough.' Steve Price points out that this source had in fact been seen and catalogued in the AFGL survey (RAFGL 5379).

16 http://spider.ipac.caltech.edu/staff/tchester/iras/no_tenth_planet_yet.htm. Tom Chester seems to suggest that a completely different object, near the galaxy Messier 31, was the origin of the rumour. Because this was not a particularly bright object and was nowhere near the ecliptic plane, it is hard to see how it could have been confused with Planet X. David Morrison has given a trenchant rebuttal of Planet Nibaru in *The Sceptical Enquirer*.

17 Walker and Rowan-Robinson 1984.

18 Soifer et al. 1986.

19 Lawrence et al. 1986.

20 de Jong et al. 1984.

21 Sanders and Mirabel 1996.
22 Sanders et al. 1988a, b.
23 Helou 1986.
24 Rowan-Robinson and Crawford 1989.
25 Pat Roche, David Aitken and their colleagues measured infrared spectra for 60 of these galaxies in the mid-infrared (8–23 microns) with a variety of ground-based telescopes, and these detailed spectra showed an even clearer separation of galaxy types than the *IRAS* colours had (Roche, Aitken, Smith and Ward 1991). Later Jim Condon and his collaborators (Condon, Huang, Yin and Thuan 1991) drew attention to the class of high radio surface brightness starbursts like Arp 220, and Andreas Efstathiou and I modelled them in terms of a much more heavily obscured starburst.
26 Rowan-Robinson et al. 1990; Strauss et al. 1992; Fisher et al. 1993, 1995; Saunders et al. 2000.
27 This is true provided the cosmological constant is zero. See Chapter 8 for more on the cosmological constant.
28 Yahil, Walker and Rowan-Robinson 1986.
29 Saunders et al. 1991. Will Saunders also recalculated the 60-micron luminosity function of *IRAS* galaxies with the much larger QDOT sample and found results that agreed well with our earlier results but greatly increased the accuracy at the low- and high-luminosity ends (Saunders et al. 1990).
30 Hacking and Houck 1987; Hacking, Houck and Condon 1987. This was confirmed by source counts over a much wider area of sky from the *IRAS* Faint Source Survey by Carol Lonsdale and collaborators in 1990 (Lonsdale et al. 1990).
31 Rowan-Robinson et al. 1991.
32 Brown and Vanden Bout 1991, 1992; Solomon, Downes and Radford 1992.
33 Phil Solomon and Paul Vanden Bout (Solomon and Vanden Bout 2005) reviewed what was known about molecular gas at high redshift in 2005 and discussed 36 galaxies in which molecular emission has been detected. Almost all of these have infrared luminosities greater than 10^{13} solar luminosities and so can be classified as 'hyperluminous'. The carbon monoxide (CO) molecule has been observed in every case, with hydrogen cyanide (HCN) being detected in some galaxies. In addition, 609-micron submillimetre emission from neutral atomic carbon, which had first been detected by Phillips and Huggins in molecular clouds in the Milky Way in 1981 (Phillips and Huggins 1981), has been detected in several of the galaxies. The mass of molecules in these hyperluminous galaxies is in the range 4 to 100 billion solar masses, and the star-formation rates are 300–5000 solar masses per year, compared with about 1 solar mass per year in our own, so these truly are monsters.
34 Cutri et al. 1994. I reviewed the properties of the known hyperluminous galaxies in 2000 (Rowan-Robinson 2000).
35 Farrah et al. 2002.
36 Efstathiou and Rowan-Robinson 1990.
37 Efstathiou and Rowan-Robinson 1991, 1995. An alternative explanation for the lack of silicate emission was that the abundance of silicon was lower in AGNs (Granato and Danese 1994).
38 Pier and Krolik 1992, 1993.
39 Silva, Granato, Bressan and Danese 1998.
40 Efstathiou, Rowan-Robinson and Siebenmorgen 2000.
41 Beichman 1987.
42 Habing et al. 1985.

43 The Dutch Low-Resolution Spectrometer yielded some interesting mid-infrared spectroscopic information on bright circumstellar dust shells and planetary nebulae (Olnon et al. 1986).
44 Beichman et al. 1986.
45 Lada and Wilking 1984.
46 Shu, Adams and Lizano 1987.

8. THE *COSMIC BACKGROUND EXPLORER* AND THE RIPPLES, THE *WILKINSON MICROWAVE ANISOTROPY PROBE* AND DARK ENERGY

1 Mather and Bosough 1996. This book gives a very detailed history of the *COBE* mission and of the interactions of the *COBE* team.
2 Mather et al. 1990.
3 Smoot et al. 1992. The full story of how the CMB fluctuations were announced is given in *The Very First Light* by Mather and Bosough. I describe the impact in the United Kingdom in my book *Ripples in the Cosmos* (Rowan-Robinson 1993). John Mather and George Smoot shared the Nobel Prize for Physics in 2006 for the *COBE* discoveries.
4 Riess et al. 1998; Perlmutter et al. 1999. The supernova evidence for dark energy was primarily a discovery of the optical band, but subsequent follow-up of high-redshift supernovae has involved a variety of infrared instruments, for example NICMOS on the *Hubble Space Telescope* (Riess et al. 2001), UKIRT and CFHT (Tonry et al. 2003; Astier et al. 2006; Cooke et al. 2009). Saul Perlmutter, Brian Schmidt and Adam Riess shared the 2011 Nobel Prize for Physics for the discovery of the accelerating expansion of the universe.
5 Hauser et al. 1998.
6 Fixsen et al. 1998.
7 Puget et al. 1996. The detectability of the background hinges on the ability to model the foreground radiation from interstellar dust in the Galaxy. Puget and his collaborators had used a new Dutch radio survey of the Galaxy for neutral hydrogen to model this. One of the surprising DIRBE results was that many regions of the sky show excess millimetre emission compared with the predictions of standard grain models. This is generally interpreted as being caused by spinning grains (Draine and Lazarian 1998) that rotate tens of billions of times a second.
8 Spergel et al. 2003, 2007. Two earlier balloon-borne experiments, BOOMERanG, which was flown in Antarctica in a large circle around the South Pole, and MAXIMA, which was flown from Palestine, Texas, also found that the universe is spatially flat (de Bernadis et al. 2000; Balbi et al. 2000).

9. GIANT GROUND-BASED NEAR-INFRARED AND SUBMILLIMETRE TELESCOPES

1 McLean, Chuter, McCaughrean and Rayner 1986. Results from the many groups working with infrared arrays at the time were presented at the conference 'Astronomy with Infrared Arrays' in Hilo, Hawaii, in 1987 (Wynn-Williams and Becklin 1987). The key collaborator in infrared array development at Santa Barbara Research Center, for both the University of Rochester group of Judith Pipher and Bill Forrest and the ROE group led by Ian McLean, was Alan Hoffman.

2 Other telescopes with mirror diameters greater than 8 metres include the Hobby–Eberley Telescope (McDonald Observatory, Texas, opened in 1997), the Large Binocular Telescope (Mount Graham, Arizona, opened 2004), the Southern African Large Telescope (South African Astronomical Observatory, opened 2005) and the Gran Telescopia Canarias (La Palma, Spain, opened 2009).

3 McLean 1994, 1997, 2008.

4 McLean et al. 1998.

5 Thompson et al. 1998.

6 Skrutskie et al. 2006.

7 Epchtein et al. 1997.

8 The CFHT Legacy Survey, carried out at the Canada-France-Hawaii Telescope between 2003 and 2009, consisting of a deep survey of 4 square degrees, a wide survey of 170 square degrees, and a 'very wide' survey of 410 square degrees of the ecliptic plane, using the MegaCam wide-field CCD camera. The deep survey is targeted at supernovae and very faint galaxies. The wide survey will study large-scale structure through weak gravitational lensing (Hoekstra et al. 2006) and clustering, and stellar populations in the galaxy. The very wide survey will survey the outer Solar System as well as stellar populations and the structure of the galaxy.

9 The UKIRT Deep Sky Survey (UKIDSS), which began in 2003 and is still continuing, is a series of five surveys covering a range of areas to different depths. The UKIDSS Ultra Deep Survey will provide the most sensitive large-scale map of the distant universe ever undertaken. The aim is to understand how and when galaxies are formed and directly test our understanding of galaxy formation. UKIDSS is searching for quasars at redshifts greater than 7 and for the nearest and faintest substellar objects (brown dwarfs) (Warren et al. 2007a; Lawrence et al. 2007).

10 Bouwens et al. 2004; Bouwens, Illingworth, Blakeslee and Franx 2006; Bunker, Stanway, Ellis and McMahon 2004.

11 Lehnert et al. 2010.

12 Mortlock et al. 2011.

13 Lilly et al. 1995.

14 Lilly et al. 1996, 1999.

15 Madau et al. 1996; Madau, Pozzetti and Dickinson 1998.

16 The *Hubble Space Telescope* went on to survey a second deep field in the Southern Hemisphere (Hubble Deep Field South), and Marjn Franx and collaborators used the European Southern Observatory's Very Large Telescope to carry out deep infrared imaging of this field and deduce the presence of a population of very red galaxies at high redshifts (Franx et al. 2003). Len Cowie and his colleagues from the University of Hawaii (Cowie, Songaila, Hu and Cohen 1996) used the infrared spectrograph on the Keck Telescope to measure redshifts of a massive sample of faint galaxies, almost 400 objects, finding redshifts ranging from 0.2 to 1.7. They also found that the star-formation rate in galaxies has been declining since redshift 1. Eric Bell and his colleagues used 2MASS observations of a large sample of over 20,000 galaxies to estimate the mass of stars in each of these galaxies (Bell, McIntosh, Katz and Weinberg 2003). They concluded that the average fraction of ordinary matter in the form of stars is no more than 10%. The rest must be in the form of hot intergalactic gas.

17 Majewski, Skrutskie, Weinberg and Ostheimer 2003; Martin et al. 2004.

18 Lacey, Baas, Townes and Beballe 1979; Lacey, Townes and Hollenbach 1982; Genzel, Crawford, Townes and Watson 1985.

19 Genzel and Townes 1987.
20 Eckart and Genzel 1997.
21 Schodel et al. 2002, 2003; Genzel et al. 2003a, b; Eisenhauer et al. 2005.
22 Ghez, Klein, Morris and Becklin 1998; Ghez et al. 2003, 2005.
23 Kirkpatrick et al. 1999, 2000; Gizis et al. 2000; Geballe et al. 2002; Burgasser et al. 2002; Reid et al. 1999.
24 Kirkpatrick et al. 1999, 2000; Gizis et al. 2000; Geballe et al. 2002; Burgasser et al. 2002; Reid et al. 1999.
25 Warren et al. 2007b.
26 Tom Phillips describes how the Caltech Submillimeter Observatory came about: 'After I returned to Bell Labs in 1976, I fell into discussion with Leighton and Neugebauer on whether and how to build the CSO. This was roughly the time frame where I invented the SIS receiver, which we put on the first OVRO dish in 1979. However, after having agreed to move from Bell Labs to Caltech in 1979, I immediately was summoned to the Division Chairman's office (Robbie Vogt) to be told that a problem had set in between the National Science Foundation (NSF) and OVRO management. There was no chance that we could start the submm dish until the OVRO interferometer was running satisfactorily and I would have to take over the management and technical progress of that project before I would be allowed to start the submm telescope. This took three years and I was not able to start the CSO construction until 1984. Luckily, the submm telescope had been constructed anyway in the 1980–84 period by Bob Leighton, who had misunderstood the NSF's instructions that he was to build three telescopes, not four. His interpretation was that he had to build four for the price of three, which he did. So we hadn't actually lost all of the three years.'
27 Holland et al. 1999.
28 Hughes et al. 1998.
29 Franceschini et al. 1991.
30 Scott et al. 2002; Ivison et al. 2002; Fox et al. 2002.
31 Coppin et al. 2006.
32 Dunne et al. 2000; Dunne and Eales 2001; Eales et al. 2000.
33 Smail, Ivison and Blain 1997; Blain, Smail, Ivison and Kneib 1999.
34 Chapman et al. 2003, 2005; Barger et al. 1999.
35 Smail, Ivison, Blain and Kneib 2002; Ivison et al. 2000; Barger, Cowie and Sanders 1999; Barger, Cowie and Richards 2000; Rowan-Robinson 2001.
36 Barger, Cowie, Mushotzky and Richards 2001; Alexander et al. 2005.
37 Solomon, Downes, Radford and Barrett 1997; Sakamoto, Okumura, Ishizuki and Scoville 1999; Solomon, Downes and Radford 1992; Greve et al. 2005.
38 Tacconi et al. 2006; Bertoldi et al. 2003.
39 Kovacs et al. 2006.
40 Cole et al. 2001; Baugh et al. 2005.
41 Holland et al. 1998; Greaves et al. 1998. The JCMT and CSO were used together as an interferometer to image protostellar disks (Lay, Carlstrom, Hills and Phillips 1994), and CSO was used to image debris disks found by *Spitzer* (Chen et al. 2005). There has also been extensive work with Nobeyama, JCMT and especially IRAM on prestellar cores and T Tauri stars in star-forming regions (Ohashi, Kawabe, Ishiguro and Hayashi 1991; Tatematus et al. 1993; Ward-Thompson, Scott, Hills and André 1994; André and Montmerle 1994; Ward-Thompson, Motte and André 1999; Motte, André and Neri 1998; Kitamura et al. 2002).

10. THE *INFRARED SPACE OBSERVATORY* AND THE *SPITZER SPACE TELESCOPE*: THE STAR-FORMATION HISTORY OF THE UNIVERSE

1 Cesarsky et al. 1996.
2 De Graauw et al. 1996.
3 Lemke et al. 1996.
4 Clegg et al. 1996.
5 Van Dishoeck 2004.
6 The lines studied by *ISO* include atomic lines (neutral oxygen at 63.2 and 145.5 microns, ionized carbon at 157.7 microns), which are key diagnostics of photon versus shock heating; rotational lines of molecular hydrogen and HD; vibrational transitions of key molecules such as water, methane, acetylene, hydrogen cyanide and carbon dioxide; the 6.2-, 7.7-, 8.6- and 11.3-micron features of polycyclic aromatic hydrocarbon material; and solid-state vibrational bands of ices, silicates, carbides, carbonates and sulfides.
7 Malfait et al. 1998; Wooden et al. 1999; Meeus et al. 2001.
8 Helou et al. 2000; Dale and Helou 2002.
9 Genzel et al. 1998; Lutz et al. 1998; Rigopoulou et al. 1999.
10 Mirabel et al. 1998.
11 Efstathiou and Rowan-Robinson 1995.
12 Haas et al. 1998.
13 Sanders et al. 1989.
14 Elbaz et al. 1999; Aussel, Cesarsky, Elbaz and Starck 1999.
15 Rowan-Robinson et al. 2004.
16 Rowan-Robinson et al. 1997.
17 Elbaz et al. 2002.
18 Werner 2006.
19 Fazio et al. 2004.
20 Rieke et al. 2004.
21 Houck et al. 2004.
22 Werner 2006.
23 Soifer, Helou and Werner 2008.
24 Perez-Gonzalez et al. 2005; LeFloc'h et al. 2005.
25 Daddi et al. 2005; Reddy et al. 2006.
26 Egami et al. 2005.
27 Bruzual and Charlot 1993; Fioc and Rocca-Volmerange 1997.
28 Lagache et al. 2004; Franceschini et al. 2010.
29 Dole et al. 2006.
30 Lacy et al. 2004.
31 Rowan-Robinson et al. 2008.
32 Spoon et al. 2006.
33 Hao et al. 2005.
34 Werner et al. 2006.
35 Wyatt 2008.
36 Habing et al. 2001.
37 Wyatt 2008.
38 Lecar, Franklin, Holman and Murray 2001.
39 Strom et al. 2005.
40 Mayor and Queloz 1995.

41 Charbonneau et al. 2005; Deming, Seagar, Richardson and Harrington 2005.
42 Murakami et al. 2007.

11. OUR SOLAR SYSTEM'S DUSTY DEBRIS DISK AND THE SEARCH FOR EXOPLANETS

1 The full statement by the International Astronomical Union can be found at http://www.iau.org/public_press/news/detail/iau0603/. My diary of this General Assembly can be found at http://astro.ic.ac.uk/~mrr/starsntides/stars9.doc.
2 Hauser et al. 1984.
3 Dermott, Nicholson, Burns and Houck 1984; Dermott and Nicholson 1989; Grogan, Dermott and Gustafson 1996.
4 Jones and Rowan-Robinson 1993. These models were further refined by Kelsall et al. (1998) using data from the *COBE* mission.
5 Wyatt 2008.
6 Lecar, Franklin, Holman and Murray 2001.
7 Gomes, Levison, Tsiganis and Morbidelli 2005.
8 Luu and Jewitt 2002.
9 Kuiper 1951. In 1930, Leonard (Leonard 1930) had speculated that Pluto would prove to be the first of many objects to be discovered beyond Neptune, and Edgeworth speculated in the 1940s that the short-period comets must originate beyond Pluto (Edgeworth 1943, 1949).
10 Luu and Jewitt 2002.
11 Tsiganis, Gomes, Morbidelli and Levison 2005; Morbidelli, Levison, Tsiganis and Gomes 2005; Gomes, Levison, Tsiganis and Morbidelli 2005. The Trojan asteroids lie on Jupiter's orbit, at the gravitationally neutral fourth and fifth Lagrangian points (see Chapter 8).
12 Mayor and Queloz 1995. Although the first confirmed detection was in 1995, there had been earlier claims which were not confirmed by independent observations. In 1988, Bruce Cambell, G. A. H. Young and S. Yang had claimed there might be a planet orbiting Gamma Cephei, but this claim was disputed for several years and not confirmed until 2003. In 1992, radio astronomers Aleksander Wolszczan and Dale Frail announced the discovery of planets around the pulsar PSR 1257+12 by studying subtle variations in the timing of the pulsating radio source. These planets are believed to have formed from the debris of the supernova explosion that formed the pulsar and have no real analogy with our solar system.
13 Chauvin et al. 2004.
14 Tinetti et al. 2007.
15 Swain, Visisht and Tinetti 2008.
16 Kalas et al. 2008.
17 Marois et al. 2008.
18 Mayor et al. 2009. The Kepler mission announced in December 2011 that they had detected five planets around a star, Kepler 20, two of which may be of approximately Earth mass.
19 Greaves 2005, 2006.

12. THE FUTURE: PIONEERING SPACE MISSIONS AND GIANT
GROUND-BASED TELESCOPES

1 Doyle, Pilbratt and Tauber 2009; Tauber et al. 2010.
2 Lamarre et al. 2010.
3 Mandolesi et al. 2010.
4 Pilbratt 2008, Pilbratt et al. 2010.
5 Griffin et al. 2009, 2010.
6 Poglitsch et al. 2009, 2010.
7 De Graauw et al. 2009, 2010.

Credits for Illustrations

Figures 1.1, 1.4, 6.2, 7.1, 7.2, 7.3, 7.4, 7.5, 7.6, 8.1, 8.3, 9.1, 10.8, 10.10, 11.4, 12.7, Plates I, III, IV, V, VI, VII, VIII, IX, XI, XII, XIII, XIV, XV, XVI, XVII, XVIII, XIX, XX, XXI, XXII: NASA; Figure 1.3: NRAO; Figure 2.1: John Herschel-Shorland; Figures 2.2, 2.3: Science Photo Library; Figure 3.1: Royal Observatory Edinburgh; Figure 4.1: D. Milon; Figure 4.3: Edith Low; Figures 4.4, 4.7: California Institute of Technology Archives; Figure 4.6: Bob Gehrz; Figure 4.8: George Rieke; Figure 4.9: *Astrophysical Journal*; Figure 4.10: Louis Allamandola; Figure 5.1: adapted from Naylor et al. (2000). Figure 5.2: Pat Daly; Figure 5.3: Ian Robson; Figure 5.4: Martin Harwit; Figure 5.5: Bill Hoffmann; Figure 6.1: Alcatel-Lucent; Figure 9.2: Gemini Observatory; Figures 10.1, 10.3, 10.4, 11.7, 12.1, 12.2, 12.3, 12.4, 12.5, 12.6, Plates XXIII, XXIV, XXV, XXVI, XXVII, XXVIII, XXIX, XXX, XXXI: ESA; Figure 10.2: Ewine van Dishoeck; Figures 10.6, 10.7: Mike Werner; Figure 10.9: Eiichi Egami; Figure 11.1: Brian May; Figure 12.8, Plate XIII: ALMA Observatory; Figures 11.3, 12.9: European Southern Observatory; Plate X: STFC; Plate XIV: Hervé Aussel.

Glossary

absolute zero the temperature at which random motions of atoms cease (−273 degrees Celsius), taken as the zero point for the Kelvin (K) scale of temperature.

absorption line dark line seen across the spectrum of a star or galaxy indicating that a particular element somewhere in the line of sight has absorbed photons.

abundances of elements the relative amounts of different elements. Nucleosynthesis in stars gradually converts hydrogen to helium and heavier elements.

active galactic nucleus (AGN) central region of a galaxy, consisting of a supermassive black hole surrounded by a disk of gas radiating at x-ray, ultraviolet, optical and infrared wavelengths.

age of universe time since the Big Bang, believed to be 13.7 billion years.

anisotropy of universe deviation of the distribution of galaxies or radiation from isotropy.

atom basic building block of matter, consisting of a nucleus of protons and neutrons surrounded by a cloud of electrons.

Big Bang the initial moment of expanding universe models, which mathematically appears as the instant at which density and other physical properties become infinite.

billion one thousand million.

black hole a region so compressed that the escape speed exceeds the speed of light, so no matter or signal can escape. The black hole mass is enclosed within an event horizon and is invisible, though we still feel its gravitational effects. Formed as the endpoint of the evolution of a very massive star (>20 times the mass of the Sun) or in the nuclei of galaxies.

blackbody a perfectly efficient absorber or emitter of radiation. It has the characteristic Planck spectrum peaking at a wavelength which depends inversely on temperature. A blackbody spectrum is the signature of a gas in which the matter and radiation are in thermal equilibrium.

brown dwarf a gaseous, self-gravitating object 10–80 times more massive than Jupiter, which is too massive to be considered a planet and of too low a mass for hydrogen burning to start and make a star.

circumstellar dust shell roughly spherical cloud of gas and dust flowing away from a star, sometimes heavily shrouding the star at visible wavelengths.

cluster of galaxies concentration of hundreds or thousands of galaxies.

cold dark matter the postulated form of nonbaryonic dark matter in which particles (e.g., the neutralino) are moving slowly in the early universe.

cosmic microwave background (CMB) background radiation, discovered at microwave wavelengths in 1965, which is a relic of the early radiation-dominated phase of the universe.

cosmic rays charged, energetic atomic particles (electrons and atomic nuclei) arriving at the Earth at speeds very close to the speed of light.

cosmological constant the parameter characterizing an additional repulsive force in the universe, whose effect increases with distance. The particle physics interpretation is that it represents the energy density of the vacuum.

cosmology study of the universe as a whole.

dark ages the period between the epoch of recombination, roughly 400,000 years after the Big Bang, until the formation of the first stars and galaxies roughly 400 million years later.

dark matter matter whose existence is inferred only from its gravitational effects.

debris disk disk of material (e.g., planetesimals, dust) left over after formation of a star; analogue of asteroid belt, Kuiper Belt and zodiacal cloud in the Solar System.

density fluctuation a region in the early universe in which the density of the universe is slightly above or below average.

detector device for detecting light.

Dewar vacuum flask storing coolant needed for operation of infrared detectors.

dipole anisotropy the effect of the Galaxy's peculiar motion through the cosmic frame, resulting in the microwave background temperature appearing slightly higher in the direction of travel and slightly cooler in the opposite direction.

Doppler shift the shift of wavelength or frequency caused by the relative motion of a source of radiation and the observer. A source of visible light moving away from us is shifted towards the red end of the spectrum, whereas a source moving towards us is shifted towards the blue end. Most galaxies appear redshifted because of the expansion of the universe.

dust small, generally submicron-sized, particles or grains of solid material, most commonly composed of silicates, graphite or amorphous carbon.

dust torus doughnut-shaped cloud of dust surrounding an active galactic nucleus.

ecliptic the great circle around the sky on which the Sun appears to move through the year as the Earth orbits around it. The ecliptic plane is the plane of the Earth's orbit. The orbits of the other planets in the solar system also lie close to this plane.

electromagnetic spectrum the whole range of types of light: x-rays, ultraviolet, visible, infrared and submillimetre light, microwaves, radio waves.

electron a basic constituent of the atom with negative charge and 1/1836 the mass of the proton. The cloud of electrons orbiting an atom determines its chemical properties.

elliptical galaxy type of galaxy consisting of an ellipsoidal distribution of old, red stars.

emission line bright line across the spectrum of a star or galaxy indicating that a particular element in the line of sight has emitted photons.

epoch of recombination the moment when the temperature of the universe drops low enough (3000 K) for electrons to combine with protons to make neutral hydrogen atoms, leaving the universe transparent to radiation.

exoplanet planet around a star other than the Sun.

expansion of universe galaxies are receding from us at a speed proportional to distance (the Hubble law), showing that we live in an expanding universe.

extinction dimming of light from stars or galaxies caused by intervening dust.

galaxy system of gas, dust and billions of stars, of which our Milky Way Galaxy is a prototype.

galaxy evolution how a galaxy changes with time as stars form out of gas and then evolve and die.

giant star star of radius much larger than the Sun. The surface temperature of the star determines its colour: red for cool stars (3000 K), blue for hot stars (10,000 K).

gravitational lens light from a background source is bent around a star, galaxy or cluster of galaxies (caused by the bending of light by matter in General Relativity), causing magnification of the source and its breakup into multiple images, arcs or a ring.

halo outer region of a galaxy, containing old stars, globular star clusters and dark matter.

heavy elements the elements from carbon onwards up the periodic table (i.e., elements other than hydrogen and the light elements).

helium second-lightest element, after hydrogen, with a nucleus consisting of two protons and two neutrons, orbited by two electrons.

H$_{II}$ region cloud of hot and ionized hydrogen.

homogeneity of the universe the universe looks the same at every location.

horizon limit to the size of the observable universe, set by the finite age of the universe and the finite speed of light.

hydrogen lightest element, consisting of a proton orbited by an electron.

interstellar medium gas and dust spread between the stars.

ionized gas gas in which most of the atoms have had some of their electrons stripped off either by collisions between atoms of gas, if the gas is hot enough, or by energetic ultraviolet or x-ray photons.

isotropy of the universe the universe looks the same in every direction.

light elements the four lightest elements, apart from hydrogen: helium, lithium, beryllium and boron.

light year distance travelled by light in one year, ten million, million kilometers.

luminosity total power output.

Messier catalogue list of 103 fuzzy or nebulous objects compiled by the eighteenth-century French comet watcher Charles Messier.

microwave background see **cosmic microwave background**.

molecular clouds dense clouds of dust and gas in which hydrogen is mainly in molecular form.

molecules a group of at least two atoms bonded together by a common electron cloud.

neutrino particle found to be emitted in radioactive β-decay of a neutron. It interacts only very weakly with other matter and has a very small mass.

neutron a heavy, uncharged subatomic particle, which with the proton is the fundamental constituent of atomic nuclei.

neutron star dead star left as a remnant of a massive star after a supernova explosion.

nuclear fusion process occurring at high densities and temperatures, for example, in the centres of stars, or in the early stages of the Big Bang, in which the nuclei of lighter atoms are fused together to make the nucleus of a heavier atom.

nucleus of galaxy inner region of a galaxy, typically a few thousand light years in diameter.

nucleosynthesis see **nuclear fusion**.

photon particle of light, carrying an energy inversely proportional to its wavelength.

proton heavy, positively charged subatomic particle, which with the neutron is the fundamental constituent of atomic nuclei.

protoplanetary disk disk of material left over after the formation of a star dense enough and massive enough to support the formation of planets.

protostar a star in the process of formation, prior to the switching on of nuclear burning at the centre.

pulsar pulsating radio source associated with a neutron star.

quasar (quasi-stellar radio source) luminous starlike object in the centre of a galaxy, caused by a supermassive black hole.

red giant star cool, luminous, very extended phase of a star's life as it exhausts its nuclear fuels.

redshift see **Doppler shift**.

Relativity, General Theory of Einstein's theory of gravity, in which space-time is curved and light is bent around masses.

spectral energy distribution the distribution of energy from a star or galaxy as a function of wavelength.

spectral line bright and dark lines across the spectrum of a star or galaxy (see **emission line**, **absorption line**).

spectrometer or spectrograph, instrument for measuring the spectrum of radiation from an astronomical source.

spectrum the spread of colours produced by passing light through a prism.

spiral galaxy galaxy containing a disk of gas, dust and newly forming massive blue stars delineating a spiral pattern.

starburst galaxy galaxy undergoing a dramatic burst of star formation, often as a result of interactions or mergers between galaxies.

star cluster aggregate of stars which have formed together. The main types are open clusters, with a few hundred young, massive stars, and globular clusters, with millions of old, lower-mass stars.

supermassive black hole black hole having a mass millions to hundreds of millions times the mass of the Sun in the nucleus of a galaxy. Responsible for active galactic nuclei.

supernova sudden explosion of a star resulting from either complete disruption of a white dwarf star (Type Ia) or the death of a massive star accompanied by the ejection of the outer parts of the star and implosion of the core to form a neutron star or black hole (Type II).

wavelength window range of wavelengths through which astronomical observations can be made from the ground.

white dwarf star a dead star left as the remnant of a star like the Sun, in which the pressure of electrons holds up the star against gravity.

zodiacal light band of visible light along the ecliptic resulting from scattering of sunlight by small particles of dust spread through the Solar System.

Further Reading

Allen, D.A. 1975. *Infrared, the New Astronomy*. Devon: Keith Reid.

Chown, M. 1996. *Afterglow of Creation*. London: Faber & Faber.

Hearnshaw, J.B. 1996. *The Measurement of Starlight: Two Centuries of Astronomical Photometry*. Cambridge: Cambridge University Press.

Leverington, D. 2000. *New Cosmic Horizons: Space Astronomy from the V2 to the Hubble Space Telescope*. Cambridge: Cambridge University Press.

Longair, M. 2006. *The Cosmic Century*. Cambridge: Cambridge University Press.

Low, F.J., Rieke, G.H. and Gehrz, R.D. 2007. 'The beginning of modern infrared astronomy'. *Annual Review of Astronomy and Astrophysics* 45, 43.

Mather, J. and Boslough, J. 1996. *The Very First Light*. New York: Basic Books.

Piazzi Smyth, C. 1858. *Teneriffe, an Astronomer's Experiment*. London: Lovell Reeve. (Reprint Cambridge: Cambridge University Press, 2010, Cambridge Library Collection.)

Rieke, G.H. 2006. *The Last of the Great Observatories: Spitzer and the Era of Faster, Better, Cheaper at NASA*. Tucson: University of Arizona Press.

Rieke, G.H. 2009. 'History of infrared telescopes and astronomy'. *Experimental Astronomy* 25, 125.

Rowan-Robinson, M. 1993. *Ripples in the Cosmos*. New York: W.H. Freeman/Spektrum.

Bibliography

I have tried to be reasonably systematic in my search of the infrared and submillimetre astronomical literature. Firstly, I have used the approximately 30 review articles in the field in the *Annual Review of Astronomy and Astrophysics* series extensively. I found Malcolm Longair's book *The Cosmic Century* invaluable in tracing the early history of infrared astronomy. I then used the ISSI Citation Index and Smithsonian ADS archive service to identify highly cited infrared and submillimetre astronomers and their key papers. I have included in the bibliography all 412 refereed papers in the field of infrared and submillimetre astronomy (excluding cosmic microwave background science) identified as having more than 200 citations in the literature as of November 2009. These papers are indicated with an asterisk. Table 1 lists the 62 astronomers with six or more papers having more than 200 citations. While citations are only an approximate indicator of impact and influence on the field, perhaps biased towards those who led the exploitation of the *IRAS*, *ISO* and *Spitzer* infrared missions, I found it satisfying that Gerry Neugebauer, whose Two Micron Survey was of such importance in launching modern infrared astronomy and whose leadership of the *IRAS* mission was so crucial, should head this table. The impact of Frank Low in pioneering far-infrared astronomy is perhaps underestimated by this metric. The impact of some names in this list is exaggerated, while other key figures, especially instrumentalists, are missing. The earliest papers in the compilation are Trumpler's interstellar dust paper of 1930 (201 citations) and Whitford's paper from 1958 on the interstellar reddening law (725 citations). Earlier pioneers from William Herschel onwards are over the citation horizon, but in ensuring that this book touches on the work of these 412 highly cited papers, among others, I hope that it reaches a certain level of comprehensiveness.

Table 1. *Highly Cited Infrared and Submillimetre Astronomers (number of infrared or submillimetre papers cited more than 200 times by November 2009)*

Gerry Neugebauer	27
Tom Soifer	21
Reinhard Genzel	20
David Sanders	17
Nick Scoville	17
Michael Rowan-Robinson	14
Rob Ivison	13
Keith Matthews	13
Eric Becklin	12
Charles Lada	12
George Rieke	12
Charles Beichman	11
Andrew Blain	11
Bruce Draine	11
Ian Smail	11
Mike Hauser	11
Xander Tielens	11
Frank Low	10
Leo Blitz	9
Thyjs de Jong	9
Andreas Eckart	9
Jay Elias	9
Fred Gillett	9
James Houck	9
Harm Habing	9
Andrew Lawrence	9
Phil Solomon	9
Erick Young	9
George Helou	8
David Hollenbach	8
Simon Lilly	8
John Mathis	8
Patrick Thaddeus	8
Len Cowie	7
Jean-Paul Kneib	7
Mark Morris	7
Frank Shu	7
Ned Wright	7
Philippe André	6

George Aumann	6
Steve Beckwith	6
Daniela Calzetti	6
Catherine Cesarsky	6
Peter Clegg	6
Martin Cohen	6
David Crampton	6
Roc Cutri	6
Thomas Dame	6
Dennis Downes	6
David Elbaz	6
Jim Emerson	6
François Hammer	6
Davy Kirkpatrick	6
Olivier Le Fèvre	6
Carol Lonsdale	6
Dieter Lutz	6
Alan Moorwood	6
Seb Oliver	6
Jean-Loup Puget	6
Stephan Price	6
Judith Young	6
Ben Zuckerman	6

* = papers with more than 200 citations by November 2009

*Aannestad, P.A. and Purcell, E.M. 1973. 'Interstellar grains'. *Astrophysical Journal* 186, 705.

*Aaronson, M., Huchra, J. and Mould, J. 1979. 'The infrared luminosity/HI velocity-width relation and its application to the distance scale'. *Astrophysical Journal* 229, 1.

*Aaronson, M. et al. 1982. 'A catalog of infrared magnitudes and HI velocity widths for nearby galaxies'. *Astrophysical Journal Supplement Series* 50, 241.

*Aaronson, M. and Mould, J. 1983. 'A distance scale from the infrared magnitude/HI velocity-width relation IV – The morphological type dependence and scatter in the relation; the distances to nearby groups'. *Astrophysical Journal* 275, 1.

*Aaronson, M. et al. 1986. 'A distance scale from the infrared magnitude/HI velocity-width relations. V – Distance moduli to 10 galaxy clusters, and positive detection of bulk supercluster motion toward the microwave anisotropy'. *Astrophysical Journal* 302, 536.

Abbott, C.J. 1924. 'Radiometer observations of stellar energy spectra'. *Astrophysical Journal* 60, 87.

*Adams, F.C., Lada, C.J. and Shu, F.H. 1987. 'Spectral evolution of young stellar objects'. *Astrophysical Journal* 312, 788.

*Adams, F.C., Shu, F.H. and Lada, C.J. 1988. 'The disks of T Tauri stars with flat infrared spectra'. *Astrophysical Journal* 326, 865.

*Adams, F.C and Shu, F.H. 1986. 'Infrared spectra of rotating protostars'. *Astrophysical Journal* 308, 836.

Alexander, D.M. et al. 2005. 'Nature of hard X-ray background sources: optical, near-infrared, submillimeter and radio properties; X-ray spectral properties of SCUBA galaxies'. *Astrophysical Journal* 632, 736.

*Allamandola, L.J., Tielens, A.G.G.M. and Barker, J.R. 1985. 'Polycyclic aromatic hydrocarbons and the unidentified infrared emission bands – Auto exhaust along the Milky Way'. *Astrophysical Journal Letters* 290, L25.

*Allamandola, L.J., Tielens, A.G.G.M. and Barker, J.R. 1989. 'Interstellar polycyclic aromatic hydrocarbons – The infrared emission bands, the excitation/emission mechanism, and the astrophysical implications'. *Astrophysical Journal Supplement Series* 71, 733.

*Allamandola, L.J., Sandford, S.A., Tielens, A.G.G.M. and Herbst, T.M. 1992. 'Infrared spectroscopy of dense clouds in the C-H stretch region – Methanol and "diamonds"'. *Astrophysical Journal* 399, 134.

Allamandola, L.J. and Hudgins, D.M. 2003. *Solid State Astrochemistry*, ed. V. Pirronello et al., NATO Science Series II, volume 120, p. 251. Berlin: Springer.

*Allen, D.A. 1973. 'Near infra-red magnitudes of 248 early-type emission-line stars and related objects'. *Monthly Notices of the Royal Astronomical Society* 161, 145.

Allen, D.A. et al. 1977. 'Optical, infrared and radio studies of AFCRL sources'. *Astrophysical Journal* 217, 108.

*Anders, E. and Zinner L. 1993. 'Interstellar grains in primitive meteorites – Diamond, silicon carbide, and graphite'. *Meteoritics* 28, 490.

*André, P., Ward-Thompson, D. and Barsony, M. 1993. 'Submillimeter continuum observations of Rho Ophiuchi A – The candidate protostar VLA 1623 and prestellar clumps'. *Astrophysical Journal* 406, 122.

*André, P. and Montmerle, T. 1994. 'From T Tauri stars to protostars: circumstellar material and young stellar objects in the rho Ophiuchi cloud'. *Astrophysical Journal* 420, 837.

*André, P., Ward-Thompson, D. and Barsony, M. 2000. 'From prestellar cores to protostars: the initial conditions of star formation'. In *Protostars and Planets IV*, ed. V. Mannings, A.P. Boss and S.S. Russell, p. 59. Tucson: University of Arizona Press.

*Angel, J.R.P. and Stockman, H.S. 1980. 'Optical and infrared polarization of active extragalactic objects'. *Annual Review of Astronomy and Astrophysics* 18, 321.

*Armandroff, T.E. and Zinn, R. 1988. 'Integrated-light spectroscopy of globular clusters at the infrared Ca II lines'. *Astronomical Journal* 96, 92.

Armitage, A. 1953. *William Herschel*. London: Nelson.

Astier, P. et al. 2006. 'The Supernova Legacy Survey: measurement of Ω_M, Ω_Λ and w from the first year data set'. *Astronomy and Astrophysics* 447, 31.

Aumann, H.H., Gillespie, C.M. and Low, F.J. 1969. 'The internal powers and effective temperatures of Jupiter and Saturn'. *Astrophysical Journal Letters* 157, 69.

*Aumann, H.H. et al. 1984. 'Discovery of a shell around Alpha Lyrae'. *Astrophysical Journal Letters* 278, L23.

*Aussel, H., Cesarsky, C.J., Elbaz, D. and Starck, J.L. 1999. 'ISOCAM observations of the Hubble Deep Field reduced with the PRETI method'. *Astronomy and Astrophysics* 342, 313.

*Bakes, E.L.O. and Tielens, A.G.G.M. 1994. 'The photoelectric heating mechanism for very small graphitic grains and polycyclic aromatic hydrocarbons'. *Astrophysical Journal* 427, 822.

Balbi, A. et al. 2000. 'Constraints on cosmological parameters from MAXIMA-1'. *Astrophysical Journal Letters* 545, L1.

*Bally, J. and Lada, C.J. 1983. 'The high-velocity molecular flows near young stellar objects'. *Astrophysical Journal* 265, 824.

*Barger, A.J. et al. 1998. 'Submillimetre-wavelength detection of dusty star-forming galaxies at high redshift'. *Nature* 394, 248.

*Barger, A.J., Cowie, L.L. and Sanders, D.B. 1999. 'Resolving the submillimeter background: the 850 micron galaxy counts'. *Astrophysical Journal Letters* 518, L5.

Barger, A.J. et al. 1999. 'Redshift distribution of submillimetre galaxies – Keck spectroscopy of SCUBA lensed galaxies'. *Astronomical Journal* 117, 2656.

*Barger, A.J., Cowie, L.L. and Richards, E.A. 2000. 'Mapping the evolution of high-redshift dusty galaxies with submillimeter observations of a radio-selected sample'. *Astronomical Journal* 119, 2092.

*Barger, A.J., Cowie, L.L., Mushotzky, R.F. and Richards, E.A. 2001. 'The nature of the hard x-ray background sources: optical, near-infrared, submillimeter, and radio properties'. *Astronomical Journal* 121, 662.

*Barlow, M.J. and Cohen, M. 1977. 'Infrared photometry and mass loss rates for OBA supergiants and Of stars'. *Astrophysical Journal* 213, 737.

*Barvainis, R. 1987. 'Hot dust and the near-infrared bump in the continuum spectra of quasars and active galactic nuclei'. *Astrophysical Journal* 320, 537.

Bastin, J.A. et al. 1964. 'Spectroscopy at extreme infra-red wavelengths. III. Astrophysical and atmospheric measurements'. *Proceedings of the Royal Society of London A* 278, 543.

*Baugh, C.M. et al. 2005. 'Can the faint submillimetre galaxies be explained in the Λ dark matter model?'*Monthly Notices of the Royal Astronomical Society* 356, 1191.

Becklin, E.E. and Neugebauer, G. 1967. 'Observations of an infrared star in the Orion nebula'. *Astrophysical Journal* 147, 799.

*Becklin, E.E. and Neugebauer, G. 1968. 'Infrared observations of the galactic center'. *Astrophysical Journal* 151, 145.

*Becklin, E.E., Neugebauer, G., Willner, S.P. and Matthews, K. 1978. 'Infrared observations of the galactic center. IV – The interstellar extinction'. *Astrophysical Journal* 220, 831.

Beckman, J.E., Bastin, J.A. and Clegg, P.E. 1969. 'Continuous spectrum of Taurus A at 1.2 mm wavelength'. *Nature* 221, 944.

*Beckwith, S., Persson, S.E., Neugebauer, G. and Becklin, E.E. 1978.'Observations of the molecular hydrogen emission from the Orion Nebula'. *Astrophysical Journal* 223, 464.

*Beckwith, S.V.W., Sargent, A., Chini, R.S. and Guesten R. 1990. 'A survey for circumstellar disks around young stellar objects'. *Astronomical Journal* 99, 924.

*Beckwith, S.V.W. and Sargent, A. 1991. 'Particle emissivity in circumstellar disks'. *Astrophysical Journal* 381, 250.

*Bedijn, P.J. 1987. 'Dust shells around Miras and OH/IR stars – Interpretation of *IRAS* and other infrared measurements'. *Astronomy and Astrophysics* 186, 136.

Beichman, C.A. et al. 1984. 'The formation of solar type stars – IRAS observations of the dark cloud Barnard 5'. *Astrophysical Journal Letters* 278, L45.

*Beichman, C.A. et al. 1986. 'Candidate solar-type protostars in nearby molecular cloud cores'. *Astrophysical Journal* 307, 337.

Beichman, C.A. 1987. 'The IRAS view of the Galaxy and the solar system'. *Annual Review of Astronomy and Astrophysics* 25, 521.

*Beichman, C.A. et al. 1988. *Infrared Astronomical Satellite (IRAS) Catalogs and Atlases. Volume 1: Explanatory Supplement*. Washington, DC: NASA.

*Bell, E.F. and de Jong, R.S. 2001. 'Stellar mass-to-light ratios and the Tully–Fisher relation'. *Astrophysical Journal* 550, 212.

*Bell, E.F., McIntosh, D.H., Katz, N. and Weinberg, M.D. 2003. 'The optical and near-infrared properties of galaxies I. Luminosity and stellar mass functions'. *Astrophysical Journal Supplement Series* 149, 289.

*Bell, E.F. et al. 2005. 'Toward an understanding of the rapid decline of the cosmic star formation rate'. *Astrophysical Journal* 625, 23.

*Benjamin, R.A. et al. 2003. 'GLIMPSE. I. An SIRTF legacy project to map the inner galaxy'. *Publications of the Astronomical Society of the Pacific* 115, 953.

*Bennett, C.L. et al. 1992. 'Preliminary separation of galactic and cosmic microwave emission for the COBE Differential Microwave Radiometer'. *Astrophysical Journal Letters* 396, L7.

Bertoldi, F. et al. 2003. 'Dust emission from the most distant quasars'. *Astronomy and Astrophysics Letters* 406, L55.

*Bertout, C., Basri, G. and Bouvier, J. 1988. 'Accretion disks around T Tauri stars'. *Astrophysical Journal* 330, 350.

*Black, J.H. and Dalgarno, A. 1976. 'Interstellar H_2 – The population of excited rotational states and the infrared response to ultraviolet radiation'. *Astrophysical Journal* 203, 132.

*Blackwell, D.E. and Sallis, M.J. 1977. 'Stellar angular diameters from infrared photometry – Application to Arcturus and other stars; with effective temperatures'. *Monthly Notices of the Royal Astronomical Society* 180, 177.

*Blain, A.W., Kneib, J.-P., Ivison, R.J. and Smail, I. 1999. 'Deep counts of submillimeter galaxies'. *Astrophysical Journal Letters* 512, L87.

*Blain, A.W., Smail, I., Ivison, R.J. and Kneib, J.-P. 1999. 'The history of star formation in dusty galaxies'. *Monthly Notices of the Royal Astronomical Society* 302, 632.

*Blain, A.W. et al. 2002. 'Submillimeter galaxies'. *Physics Reports* 369, 111.

*Blake, G.A., Sutton, E.C., Masson, C.R. and Phillips, T.G. 1987. 'Molecular abundances in OMC-1 – The chemical composition of interstellar molecular clouds and the influence of massive star formation'. *Astrophysical Journal* 315, 621.

Blitz, L. 1979. 'A study of the molecular complexes accompanying Mon OB1, Mon OB2 and CMa OB1'. PhD thesis, Columbia University.

*Blitz, L. and Shu, F.H. 1980. 'The origin and lifetime of giant molecular cloud complexes'. *Astrophysical Journal* 238, 148.

*Blitz, L., Fich, M. and Stark, A.A. 1982. 'Catalog of CO radial velocities toward galactic H II regions'. *Astrophysical Journal Supplement Series* 49, 183.

*Blitz, L. and Spergel, D.N. 1991. 'Direct evidence for a bar at the Galactic center'. *Astrophysical Journal* 379, 631.

*Blitz L. et al. 1999. 'High-velocity clouds: building blocks of the Local Group'. *Astrophysical Journal* 514, 818.

*Bloemen, J.B.G.M. et al. 1986. 'The radial distribution of galactic gamma rays. III – The distribution of cosmic rays in the Galaxy and the CO-H_2 calibration'. *Astronomy and Astrophysics* 154, 25.

*Bontemps, S., Andre, P., Terebey, S. and Cabrit S. 1996. 'Evolution of outflow activity around low-mass embedded young stellar objects'. *Astronomy and Astrophysics* 311, 858.

*Bouchet, P. et al. 1985. 'The visible and infrared extinction law and the gas-to-dust ratio in the Small Magellanic Cloud'. *Astronomy and Astrophysics* 149, 330.

*Boulanger, F. and Perault, M. 1988. 'Diffuse infared emission from the Galaxy I. – Solar neighbourhood'. *Astrophysical Journal* 330, 964.

Bouwens, R.J. et al. 2004. 'Star formation at $z \sim 6$: The Hubble Ultra Deep Parallel Fields'. *Astrophysical Journal Letters* 606, L25.

Bouwens, R.J., Illingworth, G.D., Blakeslee, J.P. and Franx, M. 2006. 'Galaxies at $z \sim 6$: the UV luminosity function and luminosity density from 506 HUDF, HUDF Parallel ACS Field and GOODS i–dropouts'. *Astrophysical Journal* 653, 53.

Boys, C.V. 1890. 'On the heat of the moon and stars'. *Proceedings of the Royal Society* 47, 480.

*Brand, J. and Blitz, L. 1993. 'The velocity field of the outer galaxy'. *Astronomy and Astrophysics* 275, 67.

*Bronfman, L. et al. 1988. 'A CO survey of the southern Milky Way – The mean radial distribution of molecular clouds within the solar circle'. *Astrophysical Journal* 324, 248.

Brown, R.L. and Vanden Bout, P.A. 1991. 'CO emission at $z = 2.2867$ in the galaxy IRAS F10214+4724'. *Astronomical Journal* 102, 1956.

Brown, R.L. and Vanden Bout, P.A. 1992. 'IRAS F1024+4724 – an extended CO emission source at $z = 2.2867$'. *Astrophysical Journal Letters* 397, L19.

*Bruzual, G. and Charlot, S. 1993. 'Spectral evolution of stellar populations using isochrone synthesis'. *Astrophysical Journal* 405, 538.

*Bruzual, G. and Charlot, S. 2003. 'Stellar population synthesis at the resolution of 2003'. *Monthly Notices of the Royal Astronomical Society* 344, 1000.

Bunker, A.J., Stanway, E.R., Ellis, R.S. and McMahon, R.G. 2004. 'The star formation rate of the Universe at $z\sim6$ from the Hubble Ultra-Deep Field'. *Monthly Notices of the Royal Astronomical Society* 355, 374.

Burbidge, E.M., Burbidge, G.R., Fowler, W.A. and Hoyle, F. 1957. 'Synthesis of the elements in stars'. *Reviews of Modern Physics* 29, 547.

Burgasser, A.J. et al. 2002. 'The spectra of T dwarfs. I. Near-infrared data and spectral classification'. *Astrophysical Journal* 564, 421.

*Burrows, A. et al. 1997. 'A nongray theory of extrasolar giant planets and brown dwarfs'. *Astrophysical Journal* 491, 856.

*Burton, M.G., Hollenbach, D.J. and Tielens, A.G.G.M. 1990. 'Line emission from clumpy photodissociation regions'. *Astrophysical Journal* 365, 620.

*Cabrit, S., Edwards, S., Strom, S.E. and Strom, K.M. 1990. 'Forbidden-line emission and infrared excesses in T Tauri stars – Evidence for accretion-driven mass loss?'. *Astrophysical Journal* 354, 687.

*Calzetti, D. 1997. 'Reddening and star formation in starburst galaxies'. *Astronomical Journal* 113, 162.

*Calzetti, D. et al. 2000. 'The dust content and opacity of actively star-forming galaxies'. *Astrophysical Journal* 533, 682.

*Calzetti, D. 2001. 'The dust opacity of star-forming galaxies'. *Publications of the Astronomical Society of the Pacific* 113, 1449.

Cameron, A.G.W. 1957. 'Nuclear reactions in stars and nucleogenesis'. *Publications of the Astronomical Society of the Pacific* 69, 201.

*Cardelli, J.A., Clayton, G.C. and Mathis, J.S. 1989. 'The relationship between infrared, optical, and ultraviolet extinction'. *Astrophysical Journal* 345, 245.

*Carilli, C.L. and Yun, M.S. 1999. 'The radio-to-submillimetre spectral index as a redshift indicator'. *Astrophysical Journal Letters* 513, L13.

Cayrel, R. and Schatzman, E. 1954. 'Sur la polarization interstellaire par des particles de graphite'. *Annales d'Astrophysique* 17, 555.

*Cesarsky, C.J. et al. 1996. 'ISOCAM in flight'. *Astronomy and Astrophysics* 316, L32.

*Chapman, S.C., Blain, A.W., Ivison, R.J. and Smail, I.R. 2003. 'A median redshift of 2.4 for galaxies bright at submillimetre wavelengths'. *Nature* 422, 695.

*Chapman, S.C., Blain, A.W., Smail, I. and Ivison, R.J. 2005. 'A redshift survey of the submillimeter galaxy population'. *Astrophysical Journal* 622, 772.

*Charbonneau, D. et al. 2005. 'Detection of thermal emission from an extrasolar planet'. *Astrophysical Journal* 626, 523.

*Chary, R. and Elbaz, D. 2001. 'Interpreting the cosmic infrared background: constraints on the evolution of the dust-enshrouded star formation rate'. *Astrophysical Journal* 556, 562.

Chauvin, G. et al. 2004. 'A giant planet candidate near a young brown dwarf. Direct VLT/NACO observations using IR wavefront sensing'. *Astronomy and Astrophysics* 425, L29.

Chen, C.H. et al. 2005. 'A Spitzer study of dusty disks around nearby, young stars'. *Astrophysical Journal* 634, 1372.

Cheung, A.C. et al. 1968. 'Detection of NH_3 molecules in the interstellar medium by their microwave emission'. *Physical Review Letters* 21, 1701.

*Chiang, E.I. and Goldreich, P. 1997. 'Spectral energy distributions of T Tauri stars with passive circumstellar disks'. *Astrophysical Journal* 490, 368.

Chown, M. 1996. *Afterglow of Creation*. London: Faber & Faber.

Clegg, P.E., Newstead, R.A. and Bastin, J.A. 1969. 'Millimetre and submillimetre astronomy'. *Philosophical Transactions of the Royal Society of London Series A* 264, 1150.

Clegg, P.E., Ade, P.A.R. and Rowan-Robinson, M. 1974. 'Narrow-band observations of galactic and extragalactic sources at 1 mm'. *Nature* 249, 530.

*Clegg, P.E. et al. 'The *ISO* long-wavelength spectrometer'. 1996. *Astronomy and Astrophysics* 315, L38.

*Clemens, D.P. and Barvainis, R. 1988. 'A catalog of small, optically selected molecular clouds – Optical, infrared, and millimeter properties'. *Astrophysical Journal Supplement Series* 68, 257.

Coblentz, W.W. 1914. 'Note on the radiation from stars'. *Publications of the Astronomical Society of the Pacific* 26, 169.

Coblentz, W.W. 1922. 'New measurements of stellar radiation'. *Astrophysical Journal* 55, 20.

*Cohen, J.G., Persson, S.E. and Frogel, J.A. 1978. 'Infrared photometry, bolometric magnitudes, and effective temperatures for giants in M3, M13, M92, and M67'. *Astrophysical Journal* 222, 165.

*Cohen, J.G., Persson, S.E., Elias, J.H. and Frogel, J.A. 1981. 'Bolometric luminosities and infrared properties of carbon stars in the Magellanic Clouds and the Galaxy'. *Astrophysical Journal* 249, 481.

*Cohen, M. and Kuhi, L.V. 1979. 'Observational studies of pre-main-sequence evolution'. *Astrophysical Journal Supplement Series* 41, 743.

*Cohen, M. et al. 1999. 'Spectral irradiance calibration in the infrared. X. A self-consistent radiometric all-sky network of absolutely calibrated stellar spectra'. *Astronomical Journal* 117, 1864.

*Cohen, R.S. et al. 1988. 'A complete CO survey of the Large Magellanic Cloud'. *Astrophysical Journal Letters* 331, L95.

*Cole, S. et al. 2001. 'The 2dF galaxy redshift survey: near-infrared galaxy luminosity functions'. *Monthly Notices of the Royal Astronomical Society* 326, 255.

*Condon, J.J., Huang, Z.-P., Yin, Q.F. and Thuan, T.X. 1991. 'Compact starbursts in ultraluminous infrared galaxies'. *Astrophysical Journal* 378, 65.

Conway Morris, S. 2003. *Life's Solution: Inevitable Humans in a Lonely Universe*. Cambridge: Cambridge University Press.

Cooke, J. et al. 2009. 'Type IIn supernovae at redshift ~2 from archival data'. *Nature* 460, 237.

Coppin, K. et al. 2006. 'The SCUBA half-degree extragalactic survey (SHADES)'. *Monthly Notices of the Royal Astronomical Society* 372, 1621.

*Cowie, L.L., Songaila, A., Hu, E.M. and Cohen, J.G. 1996. 'New insight on galaxy formation and evolution from Keck spectroscopy of the Hawaii deep fields'. *Astronomical Journal* 112, 839.

*Crovisier, J. et al. 1997. 'The spectrum of Comet Hale-Bopp (C/1995 01) observed with the Infrared Space Observatory at 2.9 AU from the Sun'. *Science* 275, 1904.

Cutri, R.M. et al. 1994. 'IRAS F15307+3252: a hyperluminous infrared galaxy at $z = 0.93$'. *Astrophysical Journal Letters* 424, L65.

*Daddi, E. et al. 2004. 'A new photometric technique for the joint selection of star-forming and passive galaxies at $1.4 < z < 2.5$'. *Astrophysical Journal* 617, 746.

Daddi, E. et al. 2005. 'Passively evolving early-type galaxies at $1.4 < z < 2.5$ in the Hubble Ultra Deep Field'. *Astrophysical Journal* 626, 680.

*Dahn, C.C. et al. 2002. 'Astrometry and photometry for cool dwarfs and brown dwarfs'. *Astronomical Journal* 124, 1170.

*Dale, D.A. et al. 2001. 'The infrared spectral energy distribution of normal star-forming galaxies'. *Astrophysical Journal* 549, 215.

*Dale, D.A. and Helou, G. 2002. 'The infrared spectral energy distribution of normal star-forming galaxies: calibration at far-infrared and submillimeter wavelengths'. *Astrophysical Journal* 576, 159.

*D'Alessio, P., Calvet, N. and Hartmann, L. 2001. 'Accretion disks around young objects. III. Grain growth'. *Astrophysical Journal* 553, 321.

*Dame, T.M., Elmegreen, B.G., Cohen, R.S. and Thaddeus, P. 1986. 'The largest molecular cloud complexes in the first galactic quadrant'. *Astrophysical Journal* 305, 892.

*Dame, T.M. et al. 1987. 'A composite CO survey of the entire Milky Way'. *Astrophysical Journal* 322, 706.

*Dame, T.M., Hartmann, D. and Thaddeus, P. 2001. 'The Milky Way in molecular clouds: a new complete CO survey'. *Astrophysical Journal* 547, 792.

*Danchi, W.C. et al. 1994. 'Characteristics of dust shells around 13 late-type stars'. *Astronomical Journal* 107, 1469.

Davidson, K. and Harwit, M. 1967. 'Infrared and radio appearance of cocoon stars'. *Astrophysical Journal* 148, 443.

Davies, J.K. et al. 1984. 'The IRAS fast-moving object search'. *Nature* 309, 315.

Davis, L., Jr. and Greenstein, J.L. 1951. 'The polarization of starlight by aligned dust grains'. *Astrophysical Journal* 114, 206.

de Bernardis, P. et al. 2000. 'A flat Universe from high resolution maps of the cosmic microwave background radiation'. *Nature* 404, 955.

*de Graauw, T. et al. 1996. 'Observing with the *ISO* short-wavelength spectrometer'. *Astronomy and Astrophysics* 315, L49.

de Graauw, T. et al. 2009. 'The Herschel-Heterodyne Instrument for the Far-Infrared (HIFI)'. *European Astronomical Society Publication Series* 34, 3.

de Graauw, T. et al. 2010. 'The Herschel-Heterodyne Instrument for the Far-Infrared (HIFI)'. *Astronomy and Astrophysics* 518, L6.

*de Grijp, M.H.K., Miley, G.K., Lub, J. and de Jong, T. 1985. 'Infrared Seyferts – a new population of active galaxies?'. *Nature* 314, 240.

*de Jong, R.S. 1996a. 'Near-infrared and optical broadband surface photometry of 86 face-on disk dominated galaxies. III. The statistics of the disk and bulge parameters'. *Astronomy and Astrophysics* 313, 45.

*de Jong, R.S. 1996b. 'Near-infrared and optical broadband surface photometry of 86 face-on disk dominated galaxies. IV. Using color profiles to study stellar and dust content of galaxies'. *Astronomy and Astrophysics* 313, 377.

*de Jong, T. et al. 1984. 'IRAS observations of Shapley–Ames galaxies'. *Astrophysical Journal Letters* 278, L67.

*de Jong, T., Klein, U., Wielebinski, R. and Wunderlich, E. 1985. 'Radio continuum and far-infrared emission from spiral galaxies – a close correlation'. *Astronomy and Astrophysics* 147, L6.

*Deming, D., Seagar, S., Richardson, L.J. and Harrington, J. 2005. 'Infrared radiation from an extrasolar planet'. *Nature* 434, 740.

Dermott, S.F., Nicholson, P.D., Burns, J.A. and Houck, J.R. 1984. 'Origin of the solar system dust bands discovered by IRAS'. *Nature* 312, 505.

Dermott, S.F. and Nicholson, P.D. 1989. '*IRAS* dust bands and the origin of the zodiacal cloud'. *Highlights of Astronomy* 8, 259.

*Desert, F.-X., Boulanger, F. and Puget, J.L. 1990. 'Interstellar dust models for extinction and emission'. *Astronomy and Astrophysics* 237, 215.

*D'Hendecourt, L.B., Allamandola, L.J. and Greenberg, J.M. 1985. 'Time dependent chemistry in dense molecular clouds. I – Grain surface reactions, gas/grain interactions and infrared spectroscopy'. *Astronomy and Astrophysics* 152, 130.

Dicke, R.H., Peebles, P.J.E., Roll, P.G. and Wilkinson, D.T. 1965. 'Cosmic black-body radiation'. *Astrophysical Journal* 142, 414.

Dole, H. et al. 2006. 'The cosmic infrared background resolved by Spitzer. Contributions of mid-infrared galaxies to the far-infrared background'. *Astronomy and Astrophysics* 451, 517.

*Downes, D., Genzel, R., Becklin, E.E. and Wynn-Williams, C.G. 1981. 'Outflow of matter in the KL Nebula – the role of IRC2'. *Astrophysical Journal* 244, 869.

*Downes, D. and Solomon, P.M. 1998. 'Rotating nuclear rings and extreme starbursts in ultraluminous galaxies'. *Astrophysical Journal* 597, 615.

Doyle, D., Pilbratt, G. and Tauber, J. 2009. 'The Herschel and Planck telescopes'. *Proceedings of the Institute of Electrical and Electronics Engineers* 97, 1403.

*Draine, B.T. and Salpeter, E.E. 1979a. 'On the physics of dust grains in hot gas'. *Astrophysical Journal* 231, 77.

*Draine, B.T. and Salpeter, E.E. 1979b. 'Destruction mechanisms for interstellar dust'. 1979b. *Astrophysical Journal* 231, 438.

*Draine, B.T. and Lee, H.M. 1984. 'Optical properties of interstellar graphite and silicate grains'. *Astrophysical Journal* 285, 89.

*Draine, B.T. 1985. 'Tabulated optical properties of graphite and silicate grains'. *Astrophysical Journal Supplement Series* 57, 587.

*Draine, B.T. and Anderson, N. 1985. 'Temperature fluctuations and infrared emission from interstellar grains'. *Astrophysical Journal* 292, 494.

*Draine, B.T. 1988. 'The discrete-dipole approximation and its application to interstellar graphite grains'. *Astrophysical Journal* 333, 848.

Draine, B.T. and Lazarian, A. 1998. 'Diffuse galactic emission from spinning dust grains'. *Astrophysical Journal Letters* 494, L19.

*Draine, B.T. 2003. 'Interstellar dust grains'. *Annual Review of Astronomy and Astrophysics* 41, 241.

*Duley, W.W. and Williams, D.A. 1981. 'The infrared spectrum of interstellar dust – surface functional groups on carbon'. *Monthly Notices of the Royal Astronomical Society* 161, 145.

*Dunne, L. et al. 2000. 'The SCUBA local universe galaxy survey – I. First measurements of the submillimetre luminosity and dust mass functions'. *Monthly Notices of the Royal Astronomical Society* 315, 115.

Dunne, L. and Eales, S. 2001. 'The SCUBA Local Universe Galaxy Survey – II. 450-micron data: evidence for cold dust in bright IRAS galaxies'. *Monthly Notices of the Royal Astronomical Society* 327, 697.

*Dwek, E. et al. 1995. 'Morphology, near-infrared luminosity, and mass of the Galactic bulge from COBE DIRBE observations'. *Astrophysical Journal* 445, 716.

*Eales, S. et al. 1999. 'The Canada-UK deep submillimeter survey: first submillimeter images, the source counts, and resolution of the background'. *Astrophysical Journal* 515, 518.

Eales, S. et al. 2000. 'The Canada-UK Deep Submillimeter survey'. *Astronomical Journal* 120, 2244.

*Eckart, A. and Genzel, R. 1997. 'Stellar proper motions in the central 0.1 pc of the Galaxy'. *Monthly Notices of the Royal Astronomical Society* 284, 576.

Eddington, A. 1925. *Internal Constitution of the Stars.* Cambridge: Cambridge University Press.

Eddy, J.A. 1972. 'Thomas A. Edison and infra-red astronomy'. *Journal of the History of Astronomy* 3, 165.

Edgeworth, K.E. 1943. 'The evolution of our planetary system'. *Journal of the British Astronomical Association* 53, 181.

Edgeworth, K.E. 1949. 'The origin and evolution of the Solar System'. *Monthly Notices of the Royal Astronomical Society* 109, 600.

Efstathiou, A. and Rowan-Robinson, M. 1990. 'Radiative transfer in axisymmetric dust clouds'. *Monthly Notices of the Royal Astronomical Society* 245, 275.

Efstathiou, A. and Rowan-Robinson, M. 1991. 'Radiative transfer in axisymmetric dust clouds II – Models of rotating protostars'. *Monthly Notices of the Royal Astronomical Society* 252, 528.

Efstathiou, A. and Rowan-Robinson, M. 1995. 'Dusty discs in active galactic nuclei'. *Monthly Notices of the Royal Astronomical Society* 273, 649.

Efstathiou, A., Rowan-Robinson, M. and Siebenmorgen, R. 2000. 'Massive star formation in galaxies: radiative transfer models of the UV to millimetre emission of starburst galaxies'. *Monthly Notices of the Royal Astronomical Society* 313, 734.

Egami, E. et al. 2005. 'Spitzer and Hubble Telescope constraints on the physical properties of the $z\sim7$ galaxy strongly lensed by A2218'. *Astrophysical Journal Letters* 618, L5.

*Eisenhauer, F. et al. 2005. 'SINFONI in the Galactic Center: young stars and infrared flares in the central light-month'. *Astrophysical Journal* 628, 246.

*Elbaz, D. et al. 1999. 'Source counts from the 15 μm ISOCAM Deep Surveys'. *Astronomy and Astrophysics* 351, L37.

*Elbaz, D. et al. 2002. 'The bulk of the cosmic infrared background resolved by ISOCAM'. *Astronomy and Astrophysics* 384, 848.

*Elias, J.H. 1978a. 'An infrared study of the Ophiuchus dark cloud'. *Astrophysical Journal* 224, 453.

*Elias, J.H. 1978b. 'A study of the Taurus dark cloud complex'. *Astrophysical Journal* 224, 857.

Elias, J.H. et al. 1978. '1 millimeter continuum observations of extragalactic objects'. *Astrophysical Journal* 220, 25.

*Elias, J.H., Frogel, J.A., Matthews, K. and Neugebauer, G. 1982. 'Infrared standard stars'. *Astronomical Journal* 87, 1029.

*Elitzur, M., Goldreich, P. and Scoville, N. 1976. 'OH-IR stars. II. A model for the 1612 MHz masers'. *Astrophysical Journal* 205, 384.

*Elmegreen, B.G. and Falgarone, E. 1996. 'A fractal origin for the mass spectrum of interstellar clouds'. *Astrophysical Journal* 471, 816.

*Elvis, M. et al. 1994. 'Atlas of quasar energy distributions'. *Astrophysical Journal Supplement Series* 95, 1.

Emerson, J.P., Jennings, R.E. and Moorwood, A.F.M. 1973. 'Far-infrared observations of HII regions from balloon altitudes'. *Astrophysical Journal* 184, 401.

*Epchtein, N. et al. 1997. 'The deep near-infrared southern sky survey (DENIS)'. *ESO Messenger* 87, 27.

*Evans, N.J. II et al. 2003. 'From molecular cores to planet-forming disks: a SIRTF Legacy Program'. *Publications of the Astronomical Society of the Pacific* 115, 965.

*Fabian, A.C. 1999. 'The obscured growth of massive black holes'. *Monthly Notices of the Royal Astronomical Society* 308, L39.

*Falgarone, E., Phillips, T.G. and Walker, C.K. 1991. 'The edges of molecular clouds – fractal boundaries and density structure'. *Astrophysical Journal* 378, 186.

Farrah, D. et al. 2002. 'Hubble Space Telescope Wide Field Camera 2 observations of hyperluminous infrared galaxies'. *Monthly Notices of the Royal Astronomical Society* 329, 605.

Fazio, G.G. et al. 1974. 'A high-resolution map of the Orion Nebula region at far-infrared wavelengths'. *Astrophysical Journal Letters* 192, L23.

*Fazio, G.G. et al. 2004. 'The Infrared Array Camera (IRAC) for the Spitzer Space Telescope'. *Astrophysical Journal Supplement Series* 154, 10.

Fellgett, P.B. 1951. 'An exploration of infra-red stellar magnitudes using the photoconductivity of lead sulphide'. *Monthly Notices of the Royal Astronomical Society* 111, 537.

*Fich, M., Blitz, L. and Stark, A.A. 1989. 'The rotation curve of the Milky Way to 2 R(0)'. *Astrophysical Journal* 342, 272.

*Figer, D.F. et al. 1999. 'Hubble Space Telescope/NICMOS observations of massive stellar clusters near the Galactic Center'. *Astrophysical Journal* 525, 750.

*Finkbeiner, D.P., Davis, M. and Schlegel, D.J. 1999. 'Extrapolation of Galactic dust emission at 100 microns to cosmic microwave background radiation frequencies using FIRAS'. *Astrophysical Journal* 524, 867.

*Fioc, M. and Rocca-Volmerange, B. 1997. 'PEGASE: a UV to NIR spectral evolution model of galaxies. Application to the calibration of bright galaxy counts'. *Astronomy and Astrophysics* 326, 950.

*Fisher, K.B. et al. 1993. 'The power spectrum of *IRAS* galaxies'. *Astrophysical Journal* 402, 42.

*Fisher, K.B. et al. 1995. 'The *IRAS* 1.2 Jy survey: redshift data'. *Astrophysical Journal Supplement Series* 100, 69.

*Fixsen, D.J. et al. 1998. 'The spectrum of the extragalactic far-infrared background from the COBE FIRAS observations'. *Astrophysical Journal* 508, 123.

Fizeau, H. and Foucault, L. 1847. 'Recherches sur les interférences des rayons calorifiques'. *Comptes Rendus de l'Académie de Sciences* 25, 447.

*Flores, H. et al. 1999. '15 micron Infrared Space Observatory observations of the 1415+52 Canada-France redshift survey field: the cosmic star formation rate as derived from deep ultraviolet, optical, mid-infrared, and radio photometry'. *Astrophysical Journal* 517, 148.

Fox, M.J. et al. 2002. 'The SCUBA 8-mJy Survey – II. Multiwavelength analysis of bright submillimetre sources'. *Monthly Notices of the Royal Astronomical Society* 331, 839.

Franceschini, A. et al. 1991. 'Galaxy counts and contributions to the background radiation from 1 micron to 1000 microns'. *Astronomy and Astrophysics Supplement* 89, 285.

Franceschini, A. et al. 2010. 'Galaxy evolution from deep multi-wavelength infrared surveys: a prelude to Herschel'. *Astronomy and Astrophysics* 517, 74.

*Franx, M. et al. 2003. 'A significant population of red, near-infrared-selected high-redshift galaxies'. *Astrophysical Journal Letters* 587, L79.

*Frogel, J.A., Persson, S.E., Matthews, K. and Aaronson, M. 1978. 'Photometric studies of composite stellar systems. I – CO and JHK observations of E and S0 galaxies'. *Astrophysical Journal* 220, 75.

*Frogel, J.A., Persson, S.E. and Cohen, J.G. 1981. 'Infrared photometry of red giants in the globular cluster 47 Tucanae'. *Astrophysical Journal* 246, 842.

*Frogel, J.A. and Whitford, A.E. 1987. 'M giants in Baade's window – infrared colors, luminosities, and implications for the stellar content of E and S0 galaxies'. *Astrophysical Journal* 320, 199.

Furniss, I., Jennings, R.E. and Moorwood, A.F.M. 1972a. 'Infrared astronomy – observations of M42, NGC 2024 and M1'. *Nature* 236, 6.

Furniss, I., Jennings, R.E. and Moorwood, A.F.M. 1972b. 'Detection of far-infrared astronomical sources'. *Astrophysical Journal Letters* 176, L105.

Furniss, I., Jennings, R.E. and Moorwood, A.F.M. 1975. *Astrophysical Journal* 202, 400.

Gaitskell, J.N., Newstead, R.A. and Bastin, J.A. 1969. 'Submillimetre solar radiation at sea level'. *Philosophical Transactions of the Royal Society of London Series A* 264, 1150.

Gautier, T.N. et al. 1984. 'IRAS images of the Galactic Centre'. *Astrophysical Journal Letters* 278, L57.

*Geballe, T.R. et al. 2002. 'Toward spectral classification of L and T dwarfs: infrared and optical spectroscopy and analysis'. *Astrophysical Journal* 564, 466.

*Gehrz, R.D. and Woolf, N.J. 1971. 'Mass loss from M stars'. *Astrophysical Journal* 165, 285.

*Gehrz, R.D., Hackwell, J.A. and Jones, T.W. 1974. 'Infrared observations of Be stars from 2.3 to 19.5 microns'. *Astrophysical Journal* 191, 675.

*Genzel, R. and Downes, D. 1977. 'H_2O in the Galaxy: sites of newly formed OB stars'. *Astronomy and Astrophysics Supplement* 30, 14.

*Genzel, R., Reid, M.J., Moran, J.M. and Downes, D. 1981. 'Proper motions and distances of H_2O maser sources. I – The outflow in Orion-KL'. *Astrophysical Journal* 244, 884.

Genzel, R., Crawford, M.K., Townes, C.H. and Watson, D.M. 1985. 'The neutral-gas disk around the Galactic Center'. *Astrophysical Journal* 297, 766.

*Genzel, R. and Townes, C.H. 1987. 'Physical conditions, dynamics, and mass distribution in the center of the Galaxy'. *Annual Review of Astronomy and Astrophysics* 25, 377.

*Genzel, R. and Stutzki, J. 1989. 'The Orion Molecular Cloud and star-forming region'. *Annual Review of Astronomy and Astrophysics* 27, 41.

*Genzel, R., Eckart, A., Ott, T. and Eisenhauer, F. 1997. 'On the nature of the dark mass in the centre of the Milky Way'. *Monthly Notices of the Royal Astronomical Society* 291, 219.

*Genzel, R. et al. 1998. 'What powers ultraluminous *IRAS* galaxies?' *Astrophysical Journal* 498, 579.

Genzel, R. and Cesarsky, C.J. 2000. 'Extragalactic results from the Infrared Space Observatory'. *Annual Review of Astronomy and Astrophysics* 38, 761.

*Genzel, R. et al. 2003a. 'The stellar cusp around the supermassive black hole in the Galactic Center'. *Astrophysical Journal* 594, 812.

*Genzel, R. et al. 2003b. 'Near-infrared flares from accreting gas around the supermassive black hole at the Galactic Centre'. *Nature* 425, 934.

Gezari, D.Y., Joyce, R.R. and Simon, M. 1973. 'Observations of the Galactic nucleus at 350 microns'. *Astrophysical Journal Letters* 179, L67.

*Ghez, A.M., Neugebauer, G. and Matthews, K. 1993. 'The multiplicity of T Tauri stars in the star forming regions Taurus-Auriga and Ophiuchus-Scorpius: a 2.2 micron speckle imaging survey'. *Astronomical Journal* 106, 2005.

Ghez, A.M., Klein, B.L., Morris, M. and Becklin, E.E. 1998. 'High proper-motion stars in the vicinity of Sagittarius A: evidence for a supermassive black hole at the center of our galaxy'. *Astrophysical Journal* 509, 678.

*Ghez, A.M. et al. 2003. 'The first measurement of spectral lines in a short-period star bound to the galaxy's central black hole: a paradox of youth'. *Astrophysical Journal Letters* 586, L127.

*Ghez, A.M. et al. 2005. 'Stellar orbits around the Galactic Center black hole'. *Astrophysical Journal* 620, 744.

*Giavalisco, M. et al. 2004. 'The great observatories origins deep survey: initial results from optical and near-infrared imaging'. *Astrophysical Journal Letters* 600, L93.

Gillett, F.C., Low, F.J. and Stein, W.A. 1967. 'Infrared observations of the planetary nebula NGC 7027'. *Astrophysical Journal Letters* 149, L97.

Gillett, F.C., Low, F.J. and Stein, W.A. 1968a. 'The spectrum of NML Cygnus from 2.8 to 5.6 microns'. *Astrophysical Journal Letters* 153, L185.

Gillett, F.C., Low, F.J. and Stein, W.A. 1968b. 'Stellar spectra from 2.8 to 14 microns'. *Astrophysical Journal* 154, 677.

*Gillett, F.C. and Forrest, W.J. 1973. 'Spectra of the Becklin-Neugebauer point source and the Kleinmann-Low nebula from 2.8 to 13.5 microns'. *Astrophysical Journal* 179, 483.

*Gillett, F.C., Forrest, W.J. and Merrill, K.M. 1973. '8–13-micron spectra of NGC 7027, BD +30 3639, and NGC 6572'. *Astrophysical Journal* 183, 87.

*Gillett, F.C. et al. 1975. 'The 8–13 micron spectra of compact H II regions'. *Astrophysical Journal* 200, 609.

Gilman, R.C. 1969. 'On the composition of interstellar grains'. *Astrophysical Journal Letters* 155, L185.

*Gizis, J.E. et al. 2000. 'New neighbors from 2MASS: activity and kinematics at the bottom of the main sequence'. *Astronomical Journal* 120, 1085.

*Goldreich, P. and Kwan, J. 1974. 'Molecular clouds'. *Astrophysical Journal* 189, 441.

*Goldreich, P. and Scoville, N. 1976. 'OH-IR stars. I – Physical properties of circumstellar envelopes'. *Astrophysical Journal* 205, 144.

Gomes, R., Levison, H.F., Tsiganis, K. and Morbidelli, A. 2005. 'Origins of the cataclysmic Late Heavy Bombardment period of the terrestrial planets'. *Nature* 435, 466.

Gould, R.J. and Harwit, M. 1963. 'Expected near infrared radiation from interstellar molecular hydrogen'. *Astronomical Journal* 137, 694.

*Granato, G.L. and Danese, L. 1994. 'Thick tori around active galactic nuclei – A comparison of model predictions with observations of the infrared continuum and silicate features. *Monthly Notices of the Royal Astronomical Society* 268, 235.

Grasdalen, G.L. and Gaustad, J.E. 1971. 'A comparison of the Two Micron Survey with the Dearborn Catalog of Faint Red Stars'. *Astronomical Journal* 76, 231.

Greaves, J. 2005. 'Disks around stars and the growth of planetary systems'. *Science* 307, 68.

Greaves, J. 2006. 'Space debris and planet detection'. *Astronomy and Geophysics* 47, 21.

Greaves, J.S. et al. 1998. 'A dust ring around Epsilon Eridani: analog to the young Solar System'. *Monthly Notices of the Royal Astronomical Society* 506, L133.

Greenberg, M. 1963. 'Interstellar grains'. *Annual Review of Astronomy and Astrophysics* 1, 267.

Greve, T.R. et al. 2005. 'An interferometric CO survey of luminous submillimetre galaxies'. *Monthly Notices of the Royal Astronomical Society* 359, 1165.

Griffin, M.J. et al. 2009. 'The SPIRE instrument'. *European Astronomical Society Publication Series* 34, 33.

Griffin, M.J. et al. 2010. 'The Herschel-SPIRE instrument and its in-flight performance'. *Astronomy and Astrophysics* 518, L3.

Grogan, K., Dermott, S.F. and Gustafson, B.A.S. 1996. 'An estimation of the interstellar contribution to the zodiacal thermal emission'. *Astrophysical Journal* 472, 812.

*Guiderdoni, B. and Rocca-Volmerange, B. 1987. 'A model of spectrophotometric evolution for high-redshift galaxies'. *Astronomy and Astrophysics* 186, 1.

*Guiderdoni, B., Hivon, E., Bouchet, F.R. and Maffei, B. 1998. 'Semi-analytic modelling of galaxy evolution in the IR/submm range'. *Monthly Notices of the Royal Astronomical Society* 295, 877.

Gush, H.P., Halpern, M. and Wishnow, E.H. 1990. *Physical Review Letters* 65, 537.

Haas, M. et al. 1998. 'On the far-infrared emission of quasars'. *Astrophysical Journal Letters* 503, L109.

Habing, H.J. et al. 1985. 'Stars in the bulge of our Galaxy detected by IRAS'. *Astronomy and Astrophysics* 152, L1.

*Habing, H.J. 1996. 'Circumstellar envelopes and asymptotic giant branch stars'. *Astronomy and Astrophysics Review* 7, 97.

Habing, H.J. et al. 2001. 'Incidence and survival of remnant disks around main-sequence stars'. *Astronomy and Astrophysics* 365, 545.

Hacking, P. and Houck, J.R. 1987. 'A very deep IRAS survey at l=97°, b=30°'. *Astrophysical Journal Supplement Series* 63, 311.

Hacking, P., Houck, J.R. and Condon, J.J. 1987. 'A very deep IRAS survey – Constraints on the evolution of starburst galaxies'. *Astrophysical Journal Letters* 316, L15.

Hall, F.F., Jr. 1964. 'An infrared stellar mapping program'. *Memoirs of the Royal Society of Liège* 9, 432.

Hall, J.S. 1949. Observations of the polarized light from stars'. *Science* 109, 166.

*Hanel, R. et al. 1981. 'Infrared observations of the Saturnian system from Voyager'. *Science* 212, 192.

Hao, L. et al. 2005. 'The detection of silicate emission from quasars at 10 and 18 microns'. *Astrophysical Journal Letters* 625, L75.

Harper, D.A., Low, F.J., Rieke, G.H. and Armstrong, K.R. 1972. 'Observations of planets, nebulae and galaxies at 350 microns'. *Astrophysical Journal Letters* 177, L21.

Hauser, M.G. et al. 1984. 'IRAS observations of the diffuse infrared background'. *Astrophysical Journal Letters* 278, L15.

*Hauser, M.G. et al. 1998. 'The COBE diffuse infrared background experiment search for the cosmic infrared background. I. Limits and detections'. *Astrophysical Journal* 508, 25.

*Hauser, M.G. and Dwek, E. 2001. 'The cosmic infrared background: measurements and implications'. *Annual Review of Astronomy and Astrophysics* 39, 249.

*Heckman, T.M., Armus, L. and Miley, G.K. 1990. 'On the nature and implications of starburst-driven galactic winds'. *Astrophysical Journal Supplement Series* 74, 833.

*Helou, G., Soifer, B.T. and Rowan-Robinson, M. 1985. 'Thermal infrared and nonthermal radio – Remarkable correlation in disks of galaxies'. *Astrophysical Journal Letters* 298, L7.

*Helou, G. 1986. 'The *IRAS* colors of normal galaxies'. *Astrophysical Journal Letters* 311, L33.

*Helou, G., Khan, I.R., Malek, L. and Boehmer, L. 1988. 'IRAS observations of galaxies in the Virgo cluster area'. *Astrophysical Journal Supplement Series* 68, 151.

Helou, G. et al. 2000. 'The mid-infrared spectra of normal galaxies'. *Astrophysical Journal* 532, 21.

Herbig, G.H. 1962. 'The properties and problems of T Tauri stars and related objects'. *Advances in Astronomy and Astrophysics* 1, 47.

Herschel, J.F.W. 1840. 'On the Chemical Action of the Rays of the solar spectrum'. *Philosophical Transactions of the Royal Society of London* 130, 1.

Herschel, W. 1800a. 'Investigation of the powers of the prismatic colours to heat and illuminate objects'. *Philosophical Transactions of the Royal Society of London* 90, 255.

Herschel, W. 1800b. 'Experiments on the refrangibility of the invisible rays of the Sun'. *Philosophical Transactions of the Royal Society of London* 90, 284.

Herschel, W. 1800c. 'Experiments on the solar, and on the terrestrial rays that occasion heat, Parts I and II'. *Philosophical Transactions of the Royal Society of London* 90, 293 and 437.

*Higdon, S.J.U. et al. 2004. 'The SMART data analysis package for the infrared spectrograph on the Spitzer Space Telescope'. *Publications of the Astronomical Society of the Pacific* 116, 975.

*Hildebrand, R.H. 1983. 'The determination of cloud masses and dust characteristics from submillimetre thermal emission'. *Quarterly Journal of the Royal Astronomical Society* 24, 267.

*Hillenbrand, L.A., Strom, S.E., Vrba, F.J. and Keene, J. 1992. 'Herbig Ae/Be stars – Intermediate-mass stars surrounded by massive circumstellar accretion disks'. *Astrophysical Journal* 397, 613.

Hiltner, W.A. 1949. 'Polarisation of light from distant stars by interstellar medium'. *Science* 109, 165.

*Ho, P.T.P. and Townes, C.H. 1983. 'Interstellar ammonia'. *Annual Review of Astronomy and Astrophysics* 21, 239.

Hoekstra, H. et al. 2006. 'First cosmic shear results from the Canada-France-Hawaii Telescope Wide Synoptic Legacy Survey'. *Astrophysical Journal* 647, 116.

Hoffmann, W.F. and Frederick, C.L. 1969. 'Far-infrared observation of the Galactic-Center region at 100 microns'. *Astrophysical Journal Letters* 159, L9.

Hoffmann, W.F., Frederick, C.L. and Emery, R.J. 1971a. '100-micron map of the Galactic-Center region'. *Astrophysical Journal Letters* 164, L23.

Hoffmann, W., Frederick, K. and Emery, R. 1971b. '100-micron survey of the Galactic-Plane'. *Astrophysical Journal Letters* 170, L89.

*Holland, W.S. et al. 1998. 'Submillimetre images of dusty debris around nearby stars'. *Nature* 392, 788.

*Holland, W.S. et al. 1999. 'SCUBA: a common-user submillimetre camera operating on the James Clerk Maxwell Telescope'. *Monthly Notices of the Royal Astronomical Society* 303, 659.

*Hollenbach, D.J., Werner, M.W. and Salpeter, E.E. 1971. 'Molecular hydrogen in HI regions'. *Astrophysical Journal* 163, 165.

*Hollenbach, D. and McKee, C.F. 1979. 'Molecule formation and infrared emission in fast interstellar shocks. I. Physical processes'. *Astrophysical Journal Supplement Series* 41, 555.

*Hollenbach, D. and McKee, C.F. 1989. 'Molecule formation and infrared emission in fast interstellar shocks. III – Results for J shocks in molecular clouds'. *Astrophysical Journal* 342, 306.

*Hollenbach, D.J., Takahashi, T. and Tielens, A.G.G.M. 1991. 'Low-density photodissociation regions'. *Astrophysical Journal* 377, 192.

Holmes, R. 2008. *The Age of Wonder.* London: Harper.

Hoskins, M. 1963. *William Herschel and the Construction of the Heavens.* London: Oldbourne.

Hoskins, M. 2011. *Discoverers of the Universe: William and Caroline Herschel.* Princeton, NJ: Princeton University Press.

Houck, J.R., Soifer, B.T., Pipher, J.L. and Harwit, M. 1971. 'Rocket-infrared four-colour photometry of the Galaxy's central region'. *Astrophysical Journal Letters* 169, L31.

Houck, J.R. et al. 1984. 'Unidentified point-sources in the IRAS Minisurvey'. *Astrophysical Journal Letters* 278, L63.

*Houck, J.R. et al. 2004. 'The Infrared Spectrograph (IRS) on the Spitzer Space Telescope'. *Astrophysical Journal Supplement Series* 154, 18.

Hoyle, F. and Wickramasinghe, N.C. 1962. 'On graphite particles as interstellar grains'. *Monthly Notices of the Royal Astronomical Society* 124, 417.

*Hu, E.M. et al. 2002. 'A redshift z=6.56 galaxy behind the cluster Abell 370'. *Astrophysical Journal Letters* 568, L75.

Huang, S.-S. 1969a. 'Transfer of radiation in circumstellar dust envelopes. I. Extreme cases. *Astrophysical Journal* 157, 835.

Huang, S.-S. 1969b. 'Transfer of radiation in circumstellar dust envelopes. II. Intermediate case'. *Astrophysical Journal* 157, 843.

Huggins, P.J. et al. 1975. 'Detection of carbon monoxide in the Large Magellanic Cloud'. *Monthly Notices of the Royal Astronomical Society* 173, 69.

Huggins, W. 1868a. 'Further observations of some of the stars and nebulae'. *Philosophical Transactions of the Royal Society of London* 158, 529.

Huggins, W. 1868b. 'Note on the heat of the stars'. *Proceedings of the Royal Society of London* 17, 309.

*Hughes, D.H. et al. 1998. 'High-redshift star formation in the Hubble Deep Field revealed by a submillimetre-wavelength survey'. *Nature* 394, 241.

*Impey, C.D. and Neugebauer, G. 1988. 'Energy distributions of blazars'. *Astronomical Journal* 95, 307.

*Indebetouw, R. et al. 2005. 'The wavelength dependence of interstellar extinction from 1.25 to 8.0 μm using GLIMPSE data'. *Astrophysical Journal* 619, 931.

*Ivison, R.J. et al. 1998. 'A hyperluminous galaxy at z=2.8 found in a deep submillimetre survey'. *Monthly Notices of the Royal Astronomical Society* 298, 583.

Ivison, R.J. et al. 2000. 'The diversity of SCUBA-selected galaxies'. *Monthly Notices of the Royal Astronomical Society* 315, 209.

*Ivison, R.J. et al. 2002. 'Deep radio imaging of the SCUBA 8-mJy survey fields: submillimetre source identifications and redshift distribution'. *Monthly Notices of the Royal Astronomical Society* 337, 11.

*Jarrett, T.H. et al. 2000. '2MASS extended source catalog: overview and algorithms'. *Astronomical Journal* 119, 2498.

*Jarrett, T.H. et al. 2003. 'The 2MASS large galaxy atlas'. *Astronomical Journal* 125, 525.

Jennings, R.E. and Moorwood, A.F.M. 1971. 'Atmospheric emission measurements with a balloon-borne Michelson interferometer'. *Applied Optics* 10, 231.

Jennings, R.E. 1986. 'History of British infrared astronomy since the Second World War'. *Quarterly Journal of the Royal Astronomical Society* 27, 4.

Johnson, H.L. 1962. 'Infrared stellar photometry'. *Astrophysical Journal* 135, 69.

*Johnson, H.L. 1965. 'Interstellar extinction in the Galaxy'. *Astrophysical Journal* 141, 923.

Johnson, H.L. 1966a. 'Infrared photometry of galaxies'. *Astrophysical Journal* 143, 187.

*Johnson, H.L. 1966b. 'Astronomical measurements in the infrared'. *Annual Review of Astronomy and Astrophysics* 4, 193.

*Johnson, H.L., Iriarte, B., Mitchell, R.I. and Wisniewskj, W.Z. 1966. 'UBVRIJKL photometry of the bright stars'. *Communications of the Lunar and Planetary Laboratory* 4, 99.

Jones, M.H. and Rowan-Robinson, M. 1993. 'A physical model of the *IRAS* zodiacal bands'. *Monthly Notices of the Royal Astronomical Society* 264, 237.

Joseph, R.D., Wade, R. and Wright, G.S. 1984. 'Detection of molecular hydrogen in two merging galaxies'. *Nature* 311, 132.

Joseph, R.D., Meikle, W.P.S., Robertson, N.A. and Wright, G.S. 1984. 'Recent star formation in interacting galaxies. I – Evidence from JHKL photometry'. *Monthly Notices of the Royal Astronomical Society* 209, 111.

Joseph, R.D. and Wright, G.S. 1985. 'Recent star formation in interacting galaxies. II – Super starburst in merging galaxies'. *Monthly Notices of the Royal Astronomical Society* 214, 87.

Joyce, R.R., Gezari, D.Y. and Simon, M. 1972. '345-micron ground-based observations of M17, M82 and Venus'. *Astrophysical Journal Letters* 171, L67.

Kalas, P. et al. 2008. 'Optical images of an extrasolar planet 25 light-years from Earth'. *Science* 322, 1345.

*Kauffmann, G. and Charlot, S. 1999. 'K-band luminosity function at z=1: a powerful constraint on galaxy formation theory'. *Monthly Notices of the Royal Astronomical Society* 297, L23.

*Kaufman, M.J., Wolfire, M.G., Hollenbach, D.J. and Luhman, M.L. 1999. 'Far-infrared and submillimeter emission from Galactic and extragalactic photodissociation regions'. *Astrophysical Journal* 527, 795.

Kelsall, T. et al. 1998. 'The COBE diffuse background experiment search for the cosmic infrared background II: model of the interplanetary dust cloud'. *Astrophysical Journal* 508, 44.

*Kennicutt, R.C., Jr., et al. 2003. 'SINGS: the SIRTF nearby galaxies survey'. *Publications of the Astronomical Society of the Pacific* 115, 928.

*Kent, S.M., Dame, T.M. and Fazio, G. 1991. 'Galactic structure from the Spacelab infrared telescope. II – Luminosity models of the Milky Way'. *Astrophysical Journal* 378, 131.

*Kenyon, S.J. and Hartmann, L. 1987. 'Spectral energy distributions of T Tauri stars – Disk flaring and limits on accretion'. *Astrophysical Journal* 323, 714.

*Kenyon, S.J. and Hartmann, L. 1995. 'Pre-main-sequence evolution in the Taurus-Auriga molecular cloud'. *Astrophysical Journal Supplement Series* 101, 117.

*Kessler, M.F. et al. 1996. 'The Infrared Space Observatory (ISO) mission'. *Astronomy and Astrophysics* 315, L27.

*Kinney, A.L. et al. 1996. 'Template ultraviolet to near-infrared spectra of star-forming galaxies and their application to K-corrections'. *Astrophysical Journal* 467, 38.

*Kirkpatrick, J.D., Henry, T.J. and McCarthy, D.W., Jr. 1991. 'A standard stellar spectral sequence in the red/near-infrared – Classes K5 to M9'. *Astrophysical Journal Supplement Series* 77, 417.

*Kirkpatrick, J.D. et al. 1999. 'Dwarfs cooler than "M": the definition of spectral type "L" using discoveries from the 2 Micron All-Sky Survey (2MASS)'. *Astrophysical Journal* 519, 802.

*Kirkpatrick, J.D. et al. 2000. '67 additional L dwarfs discovered by the Two Micron All Sky Survey'. *Astronomical Journal* 120, 447.

Kitamura, Y. et al. 2002. 'Investigation of the physical properties of protoplanetary disks around T Tauri stars by a 1 arcsecond imaging survey: evolution and diversity of the disks in their accretion stage'. *Astrophysical Journal* 581, 357.

Kleinmann, D.E. and Low, F.J. 1967. 'Discovery of an infrared nebula in Orion'. *Astrophysical Journal Letters* 149, L1.

Kleinmann, D.E. and Low, F.J. 1970a. 'Observations of infrared galaxies'. *Astrophysical Journal Letters* 159, L165.

Kleinmann, D.E. and Low, F.J. 1970b. 'Infrared observations of galaxies and the extended nucleus of M82'. *Astrophysical Journal Letters* 161, L203.

Kleinmann, S.G., Gillett, F.C. and Joyce, R.R. 1981. 'Preliminary results of the Air Force infrared sky survey'. *Annual Review of Astronomy and Astrophysics* 19, 411.

*Kleinmann, S.G. and Hall, D.N.B. 1986. 'Spectra of late-type standard stars in the region 2.0–2.5 microns'. *Astrophysical Journal Supplement Series* 62, 501.

*Knapp, G.R. et al. 1982. 'Mass loss from evolved stars. I – Observations of 17 stars in the CO(2–1) line'. *Astrophysical Journal* 252, 616.

Knapp, G.R. 1985. 'Mass-loss from evolved stars. IV – The dust-to-gas ratio in the envelopes of Mira variables and carbon stars'. *Astrophysical Journal* 293, 273.

*Knapp, G.R. and Morris, M. 1985. 'Mass loss from evolved stars. III – Mass loss rates for fifty stars from CO J = 1–0 observations'. *Astrophysical Journal* 292, 640.

*Knapp, G.R., Guhathakurta, P., Kim, D.-W. and Jura, M.A. 1989. 'Interstellar matter in early-type galaxies. I – IRAS flux densities'. *Astrophysical Journal Supplement Series* 70, 329.

*Kochanek, C.S. et al. 2001. 'The K-band galaxy luminosity function'. *Astrophysical Journal* 560, 566.

*Koorneef, J. 1983. 'Near-infrared photometry. II – Intrinsic colours and the absolute calibration from one to five microns'. *Astronomy and Astrophysics* 128, 84.

Kovacs, A. et al. 2006. 'SHARC-2 350 micron observations of distant submillimeter galaxies'. *Astrophysical Journal* 650, 592.

*Krabbe, A. et al. 1995. 'The nuclear cluster of the Milky Way: star formation and velocity dispersion in the central 0.5 parsec'. *Astrophysical Journal Letters* 447, L95.

Kuiper, G.P., Wilson, W. and Cashman, R.J. 1947. 'An infrared stellar spectrometer'. *Astrophysical Journal* 106, 243.

Kuiper, G.P. 1951. 'On the origin of the Solar System'. *Proceedings of the National Academy of Sciences of the United States of America* 37, 1.

*Kulkarni, S.R. et al. 1999. 'The afterglow, redshift and extreme energetics of the γ-ray burst of 23 January 1999'. *Nature* 398, 389.

Lacey, J.H., Baas, F., Townes, C.H. and Beballe, T.R. 1979. 'Observations of the motion and distribution of the ionized gas in the central parsec of the Galaxy'. *Astrophysical Journal Letters* 227, L17.

Lacey, J.H., Townes, C.H. and Hollenbach, D.J. 1982. 'The nature of the central parsec of the Galaxy'. *Astrophysical Journal* 262, 120.

Lacy, M. et al. 2004. 'Obscured and unobscured active galactic nuclei in the Spitzer Space Telescope First Look Survey' *Astrophysical Journal Supplement Series* 154, 166.

*Lada, C.J. and Wilking, B.A. 1984. 'The nature of the embedded population in the Rho Ophiuchi dark cloud – Mid-infrared observations'. *Astrophysical Journal* 287, 610.

*Lada, C.J. and Adams, F.C. 1992. 'Interpreting infrared color-color diagrams – Circumstellar disks around low- and intermediate-mass young stellar objects'. *Astrophysical Journal* 393, 278.

*Lada, C.J., Lada, E.A., Clemens, D.P. and Bally, J. 1994. 'Dust extinction and molecular gas in the dark cloud IC 5146'. *Astrophysical Journal* 429, 694.

*Lada, C.J. and Lada, E.A. 2003. 'Embedded clusters in molecular clouds'. *Annual Review of Astronomy and Astrophysics* 41, 57.

Lada, C.J. 2005. 'Star formation in the Galaxy: an observational overview'. *Progress of Theoretical Physics Supplement* 158, 1.

Lagache, G. et al. 2004. 'Polycyclic aromatic hydrocarbon to the infrared output energy of the universe at $z\sim2$'. *Astrophysical Journal Supplement Series* 154, 112.

Lamarre, J.-M. et al. 2010. 'Planck pre-launch status: The HFI instrument, from specification to actual performance'. *Astronomy and Astrophysics* 520, A9.

*Langer, W.D. and Penzias, A.A. 1990. 'C-12/C-13 isotope ratio across the Galaxy from observations of C-13/O-18 in molecular clouds'. *Astrophysical Journal* 357, 477.

Langley, S. 1886. 'On hitherto unrecognized wave-lengths'. *American Journal of Science* 32, 83.

Langley, S., 1900. 'The absorption lines in the infra-red spectrum of the Sun'. *Annals of the Smithsonian Astrophysical Observatory* 1, 5.

*Laor, A. and Draine, B.T. 1993. 'Spectroscopic constraints on the properties of dust in active galactic nuclei'. *Astrophysical Journal* 402, 441.

*Lawrence, A. and Elvis, M. 1982. 'Obscuration and the various kinds of Seyfert galaxies'. *Astrophysical Journal* 256, 410.

Lawrence, A. et al. 1986. 'Studies of IRAS sources at high Galactic latitudes. II – Results from a redshift survey at $b > 60°$'. *Monthly Notices of the Royal Astronomical Society* 219, 687.

*Lawrence, A. et al. 2007. 'The UKIRT Infrared Deep Sky Survey (UKIDSS)'. *Monthly Notices of the Royal Astronomical Society* 379, 1599.

Lay, O.P., Carlstrom, J.E., Hills, R.E. and Phillips, T.G. 1994. 'Protostellar accretion disks resolved with the JCMT-CSO interferometer'. *Astrophysical Journal Letters* 434, L75.

Lecar, M., Franklin, F.A., Holman, M.J. and Murray, N.J. 2001. 'Chaos in the Solar System'. *Annual Review of Astronomy and Astrophysics* 39, 581.

*Le Floc'h, E. et al. 2005. 'Infrared luminosity functions from the Chandra deep field-south: the Spitzer view on the history of dusty star formation at $0 <\sim z <\sim 1$'. *Astrophysical Journal* 632, 169.

*Leger, A. and Puget, J.L. 1984. 'Identification of the "unidentified" IR emission features of interstellar dust?' *Astronomy and Astrophysics* 137, L5.

*Leggett, S.K. 1992. 'Infrared colors of low-mass stars'. *Astrophysical Journal Supplement Series* 82, 351.

Lehnert, M.D. et al. 2010. 'Spectroscopic confirmation of a galaxy at redshift $z = 8.6$'. *Nature* 467, 940.

*Leinert, Ch. et al. 1993. 'A systematic approach for young binaries in Taurus'. *Astronomy and Astrophysics* 278, 129.

*Leitherer, C. and Heckman, T.M. 1995. 'Synthetic properties of starburst galaxies'. *Astrophysical Journal Supplement Series* 96, 9.

*Leitherer, C. et al. 1999. 'Starburst99: synthesis models for galaxies with active star formation'. *Astrophysical Journal Supplement Series* 123, 3.

*Lemke, D. et al. 1996. 'ISOPHOT – Capabilities and performance'. *Astronomy and Astrophysics* 315, L64.

Leonard, F.C. 1930. 'The new planet Pluto'. *Astronomical Society of the Pacific Leaflets* 1, 121.

*Leung, C.M. 1975. 'Radiation transport in dense interstellar dust clouds. I – Grain temperature'. *Astrophysical Journal* 199, 340.

Leung, C.M. 1976. 'Radiation transport in dense interstellar dust clouds. II – Infrared emission from molecular clouds associated with HII regions'. *Astrophysical Journal* 209, 75.

*Li, A. and Greenberg, J.M. 1997. 'A unified model of interstellar dust'. *Astronomy and Astrophysics* 323, 566.

*Li, A. and Draine, B.T. 2001. 'Infrared emission from interstellar dust. II. The diffuse interstellar medium'. *Astrophysical Journal* 554, 778.

*Lilly, S.J. and Longair, M.S. 1984. 'Stellar populations in distant radio galaxies'. *Monthly Notices of the Royal Astronomical Society* 211, 833.

*Lilly, S.J., Cowie, L.L. and Gardner, J.P. 1991. 'A deep imaging and spectroscopic survey of faint galaxies'. *Astrophysical Journal* 360, 79.

*Lilly, S.J. et al. 1995a.'The Canada–France redshift survey. I. Introduction to the survey, photometric catalogs, and surface brightness selection effects'. *Astrophysical Journal* 455, 50.

*Lilly, S.J. et al. 1995b.'The Canada-France redshift survey. VI. Evolution of the galaxy luminosity function to $z \sim 1$'. *Astrophysical Journal* 455, 108.

*Lilly, S.J., Le Fevre, O., Hammer, F. and Crampton, D. 1996. 'The Canada-France redshift survey: the luminosity density and star formation history of the universe to $z \sim 1$'. *Astrophysical Journal Letters* 460, L1.

*Lilly, S.J. et al. 1999. 'The Canada-United Kingdom deep submillimeter survey. II. First identifications, redshifts, and implications for galaxy evolution'. *Astrophysical Journal* 518, 641.

Lindblad, B. 1935. 'A condensation theory of meteoric matter and its cosmological significance'. *Nature* 135, 133.

*Lizano, S. and Shu, F.H. 1989. 'Molecular cloud cores and bimodal star formation'. *Astrophysical Journal* 342, 834.

Longair, M. 2006. *The Cosmic Century.* Cambridge: Cambridge University Press.

*Lonsdale, C.J. and Helou, G. 1985. *Cataloged Galaxies and Quasars Observed in the IRAS Survey.* Pasadena, CA: Jet Propulsion Laboratory.

Lonsdale, C.J. et al. 1990. 'Galaxy evolution and large-scale structure in the far-infrared. II – The IRAS faint source survey'. *Astrophysical Journal* 358, 60.

*Lonsdale, C.J. et al. 2003. 'SWIRE: the SIRTF wide-area infrared extragalactic survey'. *Publications of the Astronomical Society of the Pacific* 115, 897.

Low, F.J. 1961a. 'Gallium-doped germanium resistance thermometers'. *Advances in Cryogenic Engineering* 7, 514.

Low, F.J. 1961b. 'Low-temperature germanium bolometers'. *Journal of the Optical Society of America* 51, 1300.

Low, F.J. and Johnson, H.L. 1964. 'Stellar photometry at 10 microns'. *Astrophysical Journal* 139, 1130.

Low, F.J. 1965. 'The performance of thermal detection radiometers at 1.2 mm'. *Proceedings of the IEEE* 53, 516.

Low, F.J. and Johnson, H.L. 1965. 'The spectrum of 3C 273'. *Astrophysical Journal* 141, 336.

Low, F.J. and Tucker, W.H. 1968. 'Contribution of infrared galaxies to the cosmic background'. *Physical Review Letters* 21, 1538.

Low, F.J. 1970. 'The infrared-galaxy phenomenon'. *Astrophysical Journal Letters* 159, L173.

Low, F.J. and Aumann, H.H. 1970. 'Observations of Galactic and extragalactic sources between 50 and 300 microns'. *Astrophysical Journal Letters* 162, L79.

Low, F.J. and Krishna Swamy, K.S. 1970. 'Narrow-band infrared photometry of α Ori'. *Nature* 227, 1333.

*Low, F.J. et al. 1984. 'Infrared cirrus – New components of the extended infrared emission'. *Astrophysical Journal Letters* 278, L19.

Low, F.J., Rieke, G.H. and Gehrz, R.D. 2007. 'The beginning of modern infrared astronomy'. *Annual Review of Astronomy and Astrophysics* 45, 43.

Lunel, M.L. 1960. 'Recherches de photometrie stellaire dans l'infrarouge au moyen d'une cellule au sulfure de plomb'. *Annales d'Astrophysique* 23, 1.

*Lutz, D. et al. 1998. 'The nature and evolution of ultraluminous infrared galaxies: a mid-infrared spectroscopic survey'. *Astrophysical Journal Letters* 505, L103.

Luu, J.X. and Jewitt, D.C. 2002. 'Kuiper belt objects: relics from the accretion disk of the Sun'. *Annual Review of Astronomy and Astrophysics* 40, 63.

Lynds, B.T. and Wickramasinghe, N.C. 1968. 'Interstellar dust'. *Annual Review of Astronomy and Astrophysics* 6, 215.

*Madau, P. et al. 1996. 'High-redshift galaxies in the Hubble Deep Field: colour selection and star formation history to z~4'. *Monthly Notices of the Royal Astronomical Society* 283, 1388.

*Madau, P., Pozzetti, L. and Dickinson, M. 1998. 'The star formation history of galaxies'. *Astrophysical Journal* 498, 106.

*Maddalena, R.J., Morris, M., Moscowitz, J. and Thaddeus, P. 1986. 'The large system of molecular clouds in Orion and Monoceros'. *Astrophysical Journal* 303, 375.

*Magnani, L., Blitz, L. and Mundy, L. 1985. 'Molecular gas at high Galactic latitudes'. *Astrophysical Journal* 95, 402.

*Maiolino, R. and Rieke, G.H. 1995. 'Low-luminosity and obscured Seyfert nuclei in nearby galaxies'. *Astrophysical Journal* 454, 95.

*Majewski, S.R., Skrutskie, M.F., Weinberg, M.D. and Ostheimer, J.C. 2003. 'A two micron all sky survey view of the Sagittarius Dwarf Galaxy. I. Morphology of the Sagittarius core and tidal arms'. *Astrophysical Journal* 599, 1082.

*Malfait, K. et al. 1998. 'The spectrum of the young star HD 100546 observed with the Infrared Space Observatory'. *Astronomy and Astrophysics* 332, L25.

Mandolesi, N. et al. 2010. 'Planck pre-launch status: The Planck - LFI programme'. *Astronomy and Astrophysics* 520, A3.

*Maraston, C. 2005. 'Evolutionary population synthesis: models, analysis of the ingredients and application to high-z galaxies'. *Monthly Notices of the Royal Astronomical Society* 362, 799.

*Marconi, A. and Hunt, L.K. 2003. 'The relation between black hole mass, bulge mass, and infrared luminosity'. *Astrophysical Journal Letters* 589, L21.

*Marigo, P. et al. 2008. 'Evolution of asymptotic giant branch stars II. Optical to far-infrared isochrones with improved TP-AGB models'. *Astronomy and Astrophysics* 482, 883.

Marois, C. et al. 2008. 'Direct imaging of multiple planets orbiting the star HR 8799'. *Science* 322, 1348.

*Marscher, A.P. and Gear, W.K. 1985. 'Models for high-frequency radio outbursts in extragalactic sources, with application to the early 1983 millimeter-to-infrared flare of 3C 273'. *Astrophysical Journal* 298, 114.

Martin, N.F. et al. 2004. 'A dwarf galaxy remnant in Canis Major: the fossil of an in-plane accretion onto the Milky Way'. *Monthly Notices of the Royal Astronomical Society* 348, 12.

*Mather, J.C. et al. 1994. 'Measurement of the cosmic microwave background spectrum by the COBE FIRAS instrument'. *Astrophysical Journal* 420, 439.

*Mather, J.C. et al. 1999. 'Calibrator design for the COBE Far-Infrared Absolute Spectrophotometer (FIRAS)'. *Astrophysical Journal* 512, 511.

Mather, J. and Boslough, J. 1996. *The Very First Light*. New York: Basic Books.

*Mathis, J.S., Rumpl, W. and Nordsieck, K.H. 1977. 'The size distribution of interstellar grains'. *Astrophysical Journal* 217, 425.

*Mathis, J.S., Mezger, P.G. and Panagia, N. 1983. 'Interstellar radiation field and dust temperatures in the diffuse interstellar matter and in giant molecular clouds'. *Astronomy and Astrophysics* 128, 212.

*Mathis, J.S. and Whiffen, G. 1989. 'Composite interstellar grains'. *Astrophysical Journal* 341, 808.

*Mathis, J.S. 1990. 'Interstellar dust and extinction'. *Annual Review of Astronomy and Astrophysics* 28, 37.

Matsumoto, T. et al. 1988. 'The submillimeter spectrum of the cosmic background radiation'. *Astrophysical Journal* 329, 567.

Maxwell, J.C. 1864. 'A dynamical theory of the electromagnetic field'. *Philosophical Transactions of the Royal Society of London* 155, 459.

Mayor, M. and Queloz, D. 1995. 'A Jupiter-mass companion to a solar-type star'. *Nature* 378, 355.

Mayor, M. et al. 2009. 'The HARPS search for southern extra-solar planets. XIII. A system with 3 super-Earths (4.2, 6.9 and 9.2 Earth masses)'. *Astronomy and Astrophysics* 493, 639.

*McCaughrean, M.J. and Stauffer, J.R. 1994. 'High resolution near-infrared imaging of the trapezium: a stellar census'. *Astronomical Journal* 108, 1382.

McKellar, A. 1940. 'Evidence for the molecular origin of some hitherto unidentified interstellar lines'. *Publications of the Astronomical Society of the Pacific* 52, 187.

McLean, I.S., Chuter, D.C., McCaughrean, M.J. and Rayner, J.T. 1986. 'System design of a 1–5 micron IR camera for astronomy'. *Society of Photo-optical Instrumentation Engineers Journal* 627, 430.

McLean, I.S. (ed.). 1994. *Infrared Astronomy with Arrays: The Next Generation.* Dordrecht: Kluwer.

McLean, I.S. 1997. *Electronic Imaging in Astronomy.* New York: Springer (reprinted 2008).

*McLean, I.S. et al. 1998. 'Design and development of NIRSPEC: a near-infrared echelle spectrograph for the Keck II telescope'. *Proceedings of the Society of Photo-optical Instrumentation Engineers* 3354, 566.

*Meeus, G. et al. 2001. 'ISO spectroscopy of circumstellar dust in 14 Herbig Ae/Be systems: towards an understanding of dust processing'. *Astronomy and Astrophysics* 365, 476.

Mendoza, E.E. 1966. 'Infrared photometry of T Tauri stars and related objects'. *Astrophysical Journal* 143, 1010.

Mendoza, E.E. 1968. 'Infrared excesses in T Tauri stars and related objects'. *Astrophysical Journal* 151, 977.

Menzel, D.H., Coblentz, W.W. and Lampland, C.O. 1926. 'Planetary temperatures derived from water-cell transmissions'. *Astrophysical Journal* 63, 177.

*Meurer, G.R. et al. 1997. 'The panchromatic starburst intensity limits at low and high redshift'. *Astronomical Journal* 114, 54.

*Meurer, G.R., Heckman, T.M. and Calzetti, D. 1999. 'Dust absorption and the ultraviolet luminosity density at $z\sim3$ as calibrated by local starburst galaxies'. *Astrophysical Journal* 521, 64.

*Meyer, M.R., Calvet, N. and Hillenbrand, L.A. 1997. 'Intrinsic near-infrared excess of T Tauri stars: understanding the classical T Tauri star locus'. *Astronomical Journal* 114, 288.

Mezger, P.G. and Smith, L. 1977. 'Radio observations related to star formation'. In *Star Formation*, IAU Symposium 75, ed. T. de Jong and A. Maeder, p.133. Dordrecht: Reidel.

*Mezger, P.G., Mathis, J.S. and Panagia, N. 1982. 'The origin of the diffuse galactic far infrared and sub-millimeter emission'. *Astronomy and Astrophysics* 105, 372.

Miley, G. et al. 1984. 'A 25-micron component in 3C 390.3'. *Astrophysical Journal Letters* 278, L79.

Mirabel, I.F. et al. 1998. 'The dark side of star formation in the Antennae galaxies'. *Astronomy and Astrophysics* 333, L1.

Morbidelli, A., Levison, H.F., Tsiganis, K. and Gomes, R. 2005. 'Chaotic capture of Jupiter's Trojan asteroids in the early Solar System'. *Nature* 435, 462.

Moroz, V. 1961. 'An attempt to measure the infrared radiation of the Galactic nucleus'. *Astronomicheskii Zhurnal* 38, 487.

Moroz, V. 1963. 'Radiation emission from the Orion Nebula in the 0.85–1.7 micron wavelength region'. *Astronomicheskii Zhurnal* 40, 788.

Morris, M. and Rickard, L.J. 1982. 'Molecular clouds in galaxies'. *Annual Review of Astronomy and Astrophysics* 20, 517.

*Morris, M. 1987. 'Mechanisms for mass loss from cool stars'. *Publications of the Astronomical Society of the Pacific* 99, 1115.

*Morris, M. and Serabyn, E. 1996. 'The Galactic Center environment'. *Annual Review of Astronomy and Astrophysics* 34, 645.

Mortlock, D.J. et al. 2011. 'A luminous quasar at a redshift of $z = 7.085$'. *Nature* 474, 616.

*Motte, F., André, P. and Neri, R. 1998. 'The initial conditions of star formation in the Rho Ophiuchi main cloud: wide-field millimeter continuum mapping'. *Astronomy and Astrophysics* 336, 150.

Muelner, K. and Weiss, D. 1973. 'Balloon measurements of the far-infrared background radiation'. *Physical Review D* 7, 326.

Murakami, H. et al. 2007. 'The infrared astronomical mission AKARI'. *Publications of the Astronomical Society of Japan* 59, S369.

Murdock, T.L. and Price, S.D. 1985. 'Infrared measurements of zodiacal light'. *Astronomical Journal* 90, 375.

*Myers, P.C. et al. 1987. 'Near-infrared and optical observations of IRAS sources in and near dense cores'. *Astrophysical Journal* 319, 340.

*Nakajima, T. et al. 1995. 'Discovery of a cool brown dwarf'. *Nature* 378, 463.

Naylor, D.A. et al. 2000. 'Atmospheric transmission at submillimetre wavelengths from Mauna Kea'. *Monthly Notices of the Royal Astronomical Society* 315, 622.

Neugebauer, G., Martz, D.E. and Leighton, R.B. 1965. 'Observations of extremely cool stars'. *Astrophysical Journal* 142, 399.

*Neugebauer, G. and Leighton, R.B. 1969. *The Two Micron Survey: A Preliminary Catalogue*, NASA SP-3047. Washington, DC: NASA.

Neugebauer, G., Becklin, E.E. and Hyland, A.R. 1971. 'Infrared sources of radiation'. *Annual Review of Astronomy and Astrophysics* 9, 67.

*Neugebauer, G., Oke, J.B., Becklin, E.E. and Matthews, K. 1979. 'Absolute spectral energy distribution of quasi-stellar objects from 0.3 to 10 microns'. *Astrophysical Journal* 230, 79.

*Neugebauer, G. et al. 1984. 'The Infrared Astronomical Satellite (IRAS) mission'. *Astrophysical Journal Letters* 278, L1.

Neugebauer, G., Miley, G.K., Soifer, B.T. and Clegg, P.A. 1986. 'Quasars observed by the Infrared Astronomical Satellite'. *Astrophysical Journal* 308, 815.

*Neugebauer, G. et al. 1987. 'Continuum energy distributions of quasars in the Palomar–Green Survey'. *Astrophysical Journal Supplement Series* 63, 615.

Ney, E.P. and Allen, D.A. 1969. 'The infrared sources in the Trapezium region of M42'. *Astrophysical Journal* 155, 193.

Nichols, E.F. 1901. 'On the heat radiation of Arcturus, Vega, Jupiter, and Saturn'. *Astrophysical Journal* 13, 101.

*Norman, C. and Scoville, N. 1988. 'The evolution of starburst galaxies to active galactic nuclei'. *Astrophysical Journal* 332, 124.

Ohashi, N., Kawabe, R., Ishiguro, M. and Hayashi, M. 1991. 'Molecular cloud cores in the Orion A cloud I'. *Astronomical Journal* 102, 2054.

Oke, J.B., Neugebauer, G., and Becklin, E.E. 1970. 'Absolute spectral energy distribution of quasi-stellar objects from 0.3 to 2.2 microns'. *Astrophysical Journal* 159, 341.

*Olnon, F.M. et al. 1986. 'IRAS catalogues and atlases – Atlas of low-resolution spectra'. *Astronomy and Astrophysics Supplement* 65, 607.

Olthof, H. and van Duinen, R. 1973. 'Two colour far infrared photometry of some Galactic HII regions'. *Astronomy and Astrophysics* 29, 315.

Oort, J.H., and van der Hulst, H.C. 1946. 'Gas and smoke in interstellar space'. *Bulletin of the Astronomical Institutes of the Netherlands* 10, 187.

*Ossenkopf, V. and Henning, Th. 1994. 'Dust opacities for protostellar cores'. *Astronomy and Astrophysics* 291, 943.

*Osterbrock, D.E. 1974. *Astrophysics of Gaseous Nebulae*. San Francisco: W.H. Freeman.

*Osterloh, M. and Beckwith, S.V.W. 1995. 'Millimetre-wave continuum measurements of young stars'. *Astrophysical Journal* 439, 288.

Ouchi, M. et al. 2009. 'Large area survey for $z = 7$ galaxies in SDF and GOODS-N: implications for galaxy formation and cosmic reionization'. *Astrophysical Journal* 696, 1164.

Papovich, C. et al. 2006. 'Spitzer observations of massive, red galaxies at high redshift'. *Astrophysical Journal* 640, 92.

Park, W.M., Vickers, D.G. and Clegg, P.E. 1970. 'Submillimeter radiation from the Orion Nebula'. *Astronomy and Astrophysics* 5, 325.

Parsons, L., 4th Earl of Rosse. 1873. 'On the radiation of heat from the moon, the law of its absorption by our atmosphere, and of its variation in amount with her phases'. *Proceedings of the Royal Society of London* 21, 241.

Peebles, P.J.E., Page, L.A., Jr. and Partridge, R.B. 2010. *Finding the Big Bang*. Cambridge: Cambridge University Press.

*Pei, Y.C., Fall, S.M. and Hauser, M.G. 1999. 'Cosmic histories of stars, gas, heavy elements, and dust in galaxies'. *Astrophysical Journal* 522, 604.

*Pendleton, Y.J. et al. 1994. 'Near-infrared absorption spectroscopy of interstellar hydrocarbon grains'. *Astrophysical Journal* 437, 683.

*Penzias, A.A. and Wilson, R.W. 1965a. 'A measurement of excess antenna temperature at 4080 Mc/s'. *Astrophysical Journal* 142, 419.

Penzias, A.A. and Wilson, R.W. 1965b. 'Measurement of the flux density of Cas A at 4080 Mc/c'. *Astrophysical Journal* 142, 1149.

Penzias, A.A., Jefferts, K.B. and Wilson, R.W. 1971. 'Interstellar $^{12}C^{16}O$, $^{13}C^{16}O$ and $^{12}C^{18}O$'. *Astrophysical Journal* 165, 229.

Penzias, A.A., Solomon, P.M., Wilson, R.W. and Jefferts, K.B. 1971. 'Interstellar carbon monosulfide'. *Astrophysical Journal Letters* 168, L53.

*Pérez-González, P.G. et al. 2005. 'Spitzer view on the evolution of star-forming galaxies from $z = 0$ to $z \sim 3$'. *Astrophysical Journal* 630, 82.

Perlmutter, S. et al. 1999. 'Measurements of Ω and Λ from 42 high-redshift supernovae'. *Astrophysical Journal* 517, 565.

*Persson, S.E. et al. 1998. 'A new system of faint near-infrared standard stars'. *Astronomical Journal* 116, 2475.

*Pettini, M. et al. 1998. 'Infrared observations of nebular emission lines from galaxies at $z \sim= 3$'. *Astrophysical Journal* 508, 539.

Pettit, E. and Nicholson, S.B. 1928. 'Stellar radiation measurements'. *Astrophysical Journal* 68, 279.

Pettit, E. and Nicholson, S.B. 1930. 'Lunar radiation and temperature'. *Astrophysical Journal* 71, 102.

Pettit, E. and Nicholson, S.B. 1933. 'Measurements of the radiation from variable stars'. *Astrophysical Journal* 78, 320.

Pettit, E. and Nicholson, S.B. 1935. 'Lunar radiation as related to phase'. *Astrophysical Journal* 81, 17.

Pettit, E. and Nicholson, S.B. 1936. 'Radiation from the planet Mercury'. *Astrophysical Journal* 83, 84.

Pettit, E. and Nicholson, S.B. 1940. 'Radiation measurements on the eclipsed Moon'. *Astrophysical Journal* 91, 408.

Phillips, T.G. and Jefferts, K.B. 1973. 'A low temperature heterodyne receiver for millimeter wave astronomy'. *Review of Scientific Instruments* 44, 1009.

Phillips, T.G., Jefferts, K.B., and Wannier, P.G. 1973. 'Observation of the J=2 to J=1 transition of interstellar CO at 1.3 millimeters'. *Astrophysical Journal Letters* 186, L19.

Phillips, T. and Rowan-Robinson, M. 1978. *New Scientist* 69, 170.

Phillips, T.G. and Huggins, P.J. 1981. 'Abundance of atomic carbon (CI) in dense interstellar clouds'. *Astrophysical Journal* 251, 533.

*Pickett, H.M. et al. 1998. 'Submillimeter, millimeter and microwave spectral line catalog'. *Journal of Quantitative Spectroscopy and Radiative Transfer* 60, 883.

*Pier, E.A. and Krolik, J.H. 1992. 'Infrared spectra of obscuring dust tori around active galactic nuclei I: Calculational method and basic trends'. *Astrophysical Journal* 401, 99.

*Pier, E.A. and Krolik, J.H. 1993. 'Infrared spectra of obscuring dust tori around active galactic nuclei II: Comparison with observations'. *Astrophysical Journal* 418, 673.

Pilbratt, G.L. 2008. 'Herschel mission overview and key programmes'. *Proceedings of the Society of Photo-optical Instrumentation Engineers* 7010, 1.

Pilbratt, G.L. et al. 2010. 'Hershel space observatory. An ESA facility for farinfrared and submillimetre astronomy'. *Astronomy and Astrophysics* 518, L1.

Pipher, J. 1971. 'Rocket submillimeter observations of the Galaxy and background'. PhD thesis, Cornell University.

Platt, J.R. 1956. 'On the optical properties of interstellar dust'. *Astrophysical Journal* 123, 486.

*Poggianti, B.M. 1997. 'K and evolutionary corrections from UV to IR'. *Astronomy and Astrophysics Supplement* 122, 399.

Poglitsch, A. et al. 2009. 'The PACS instrument'. *European Astronomical Society Publication Series* 34, 43.

Poglitsch, A. et al. 2010. 'The Photodetector Array Camera and Spectrometer (PACS) on the Herschel Space Observatory'. *Astronomy and Astrophysics* 518, L2.

*Pollack, J.B. et al. 1994. 'Composition and radiative properties of grains in molecular clouds and accretion disks'. *Astrophysical Journal* 421, 615.

*Pottasch, S.R. 1984. 'Planetary nebulae – A study of late stages of stellar evolution'. *Astrophyscs and Space Science Library* 107, 335.

Pouillet, C.-S. 1838. 'Mémoire sur le chaleur solaire'. *Comptes Rendus de l'Académie des Sciences* 7, 24.

*Price, S.D. and Walker, R.G. 1976. 'The AFGL four colour infrared sky survey: catalog of observations at 3.2, 11.0, 19.8 and 27.4 microns'. *Environmental Research Papers, Hanscomb AFB, AFGL*.

*Price, S.D. et al. 2001. 'Midcourse space experiment survey of the Galactic Plane'. *Astronomical Journal* 121, 2819.

Price, S.D. 2009. 'Infrared sky surveys'. *Space Science Reviews* 142, 233.

Puget, J.-L., Leger, A. and Boulanger, F. 1985. 'Contribution of large polycyclic aromatic molecules to the infrared emission of the interstellar medium'. *Astronomy and Astrophysics* 142, L19.

*Puget, J. L. and Leger, A. 1989. 'A new component of the interstellar matter – Small grains and large aromatic molecules'. *Annual Review of Astronomy and Astrophysics* 27, 161.

*Puget, J.-L. et al. 1996. 'Tentative detection of a cosmic far-infrared background with COBE'. *Astronomy and Astrophysics* 308, L5.

*Rayner, J.T. et al. 2003. 'SpeX: a medium-resolution 0.8–5.5 micron spectrograph and imager for the NASA Infrared Telescope Facility'. *Publications of the Astronomical Society of the Pacific* 115, 362.

*Reach, W.T. et al. 2005. 'Absolute calibration of the Infrared Array Camera on the Spitzer Space Telescope'. *Publications of the Astronomical Society of the Pacific* 117, 978.

Reddy, N.A. et al. 2005. 'A census of optical and near-infrared selected star-forming and passively evolving galaxies at redshift $z{\sim}2$'. *Astrophysical Journal* 633, 748.

Rees, M.J., Silk, J.I., Werner, M.W. and Wickramasinghe, N.C. 1969. 'Infrared radiation from dust in Seyfert galaxies'. *Nature* 223, 788.

Reid, I.N. et al. 1999. 'L dwarfs and the substellar mass function'. *Astrophysical Journal* 521, 613.

*Reipurth, B. and Bally, J. 2001. Herbig-Haro flows: probes of early stellar evolution'. *Annual Review of Astronomy and Astrophysics* 39, 403.

*Rice, W. et al. 1988. 'A catalog of *IRAS* observations of large optical galaxies'. *Astrophysical Journal Supplement* 68, 91.

Rickard, L.J. et al. 1975. 'Detection of extragalactic carbon monoxide at millimeter wavelengths'. *Astrophysical Journal* 199, 175.

*Ridgway, S.T., Joyce, R.R., White, N.M. and Wing, R.F. 1980. 'Effective temperatures of late-type stars – The field giants from K0 to M6'. *Astrophysical Journal* 235, 126.

*Rieke, G.H. and Low, F.J. 1972. 'Infrared photometry of extragalactic sources'. *Astrophysical Journal Letters* 176, L95.

Rieke, G.H. and Low, F.J. 1975. 'Measurements of galactic nuclei at 34 microns'. *Astrophysical Journal Letters* 200, L67.

*Rieke, G.H. 1978. 'The infrared emission of Seyfert galaxies'. *Astrophysical Journal* 226, 550.

Rieke, G.H. and Lebofsky, M.J. 1979. 'Infrared emission of extragalactic sources'. *Annual Review of Astronomy and Astrophysics* 17, 477.

*Rieke, G.H. et al. 1980. 'The nature of the nuclear sources in M82 and NGC 253'. *Astrophysical Journal* 238, 24.

*Rieke, G.H. and Lebofsky, M.J. 1985. 'The interstellar extinction law from 1 to 13 microns'. *Astrophysical Journal* 288, 618.

*Rieke, G.H. et al. 1985. '10^{12} solar luminosity starbursts and shocked molecular hydrogen in the colliding galaxies ARP 220 (= IC 4553) and NGC 6240'. *Astrophysical Journal* 290, 116.

*Rieke, G.H., Loken, K., Rieke, M.J. and Tamblyn, P. 1993. 'Starburst modeling of M82 – Test case for a biased initial mass function'. *Astrophysical Journal* 412, 99.

*Rieke, G.H. et al. 2004. 'The Multiband Imaging Photometer for Spitzer (MIPS)'. *Astrophysical Journal Supplement Series* 154, 25.

Rieke, G.H. 2009. 'History of infrared telescopes and astronomy'. *Experimental Astronomy* 25, 125.

Riess, A. et al. 1998. 'Observational evidence from supernovae for an accelerating universe and a cosmological constant'. *Astronomical Journal* 116, 1009.

*Riess, A. et al. 2001. 'The farthest known supernova: support for an accelerating universe and a glimpse of the epoch of deceleration'. *Astrophysical Journal* 560, 49.

Rigopoulou, D., Lawrence, A. and Rowan-Robinson, M. 1996. 'Multiwavelength energy distributions of ultraluminous infrared galaxies – I. Submillimetre and x-ray observations'. *Monthly Notices of the Royal Astronomical Society* 278, 1049.

*Rigopoulou, D. et al. 1999. 'A large mid-infrared spectroscopic and near-infrared imaging survey of ultraluminous infrared galaxies: their nature and evolution'. *Astronomical Journal* 118, 2625.

Ring, J. 1969. 'Infra-red and microwave astronomy'. *Journal of Optics Technology* 1, 275.

Robson, E.I. et al. 1974. 'Spectrum of the cosmic background radiation between 3 mm and 800 microns'. *Nature* 251, 591.

*Roche, P.F., Aitken, D.K., Smith, C.H. and Ward, M.J. 1991. 'An atlas of mid-infrared spectra of galaxy nuclei'. *Monthly Notices of the Royal Astronomical Society* 248, 606.

Roll, P.G. and Wilkinson, D.T. 1966. 'Cosmic background radiation at 3.2 cm – support for cosmic black-body radiation'. *Physical Review Letters* 16, 405.

Rowan-Robinson, M. 1980. 'Radiative transfer in dust clouds. I – Hot-centred clouds associated with regions of massive star formation'. *Astrophysical Journal Supplement Series* 44, 403.

Rowan-Robinson, M. and Harris, S. 1983. 'Radiative transfer in dust clouds. III – Circumstellar dust shells around late M giants and supergiants'. *Monthly Notices of the Royal Astronomical Society* 202, 767.

Rowan-Robinson, M. et al. 1984. 'The IRAS Minisurvey'. *Astrophysical Journal Letters* 278, L7.

*Rowan-Robinson, M. and Crawford, J. 1989. 'Models for infrared emission from IRAS galaxies'. *Monthly Notices of the Royal Astronomical Society* 238, 523.

Rowan-Robinson, M. et al. 1990. 'A sparse-sampled redshift survey of IRAS galaxies. I – The convergence of the IRAS dipole and the origin of our motion with respect to the microwave background'. *Monthly Notices of the Royal Astronomical Society* 247, 1.

*Rowan-Robinson, M. et al. 1991. 'A high-redshift IRAS galaxy with huge luminosity – Hidden quasar or protogalaxy?' *Nature* 351, 719.

*Rowan-Robinson, M. et al. 1997. 'Observations of the Hubble Deep Field with the Infrared Space Observatory – V. Spectral energy distributions, starburst models and star formation history'. *Monthly Notices of the Royal Astronomical Society* 289, 490.

Rowan-Robinson, M. 2000. 'Hyperluminous infrared galaxies'. *Monthly Notices of the Royal Astronomical Society* 316, 885.

Rowan-Robinson, M. et al. 2004. 'The European Large-Area ISO Survey (ELAIS): the final band-merged catalogue'. *Monthly Notices of the Royal Astronomical Society* 351, 1290.

Rowan-Robinson, M. et al. 2008. 'Photometric redshifts in the SWIRE survey'. *Monthly Notices of the Royal Astronomical Society* 386, 697.

Sakamoto, K., Okumura, S.K., Ishizuki, S. and Scoville, N.Z. 1999. 'CO images of the central regions of 20 nearby spiral galaxies'. *Astrophysical Journal Supplement Series* 124, 403.

Sandage, A.R., Becklin, E.E. and Neugebauer, G. 1969. 'UBVRIHKL photometry of the central region of M31'. *Astrophysical Journal* 157, 55.

*Sanders, D.B., Solomon, P.M. and Scoville, N.Z. 1984. 'Giant molecular clouds in the Galaxy. I – The axisymmetric distribution of H2'. *Astrophysical Journal* 276, 182.

*Sanders, D.B. and Mirabel, I.F. 1985. 'CO detections and *IRAS* observations of bright radio spiral galaxies at cz equal or less than 9000 kilometers per second'. *Astrophysical Journal Letters* 298, L31.

*Sanders, D.B., Scoville, N.Z. and Solomon, P.M. 1985. 'Giant molecular clouds in the Galaxy. II – Characteristics of discrete features'. *Astrophysical Journal* 289, 373.

*Sanders, D.B. et al. 1988a. 'Ultraluminous infrared galaxies and the origin of quasars'. *Astrophysical Journal* 325, 74.

*Sanders, D.B. et al. 1988b. 'Warm ultraluminous galaxies in the *IRAS* survey – The transition from galaxy to quasar?' *Astrophysical Journal Letters* 328, L35.

*Sanders, D.B. et al. 1989. 'Continuum energy distribution of quasars – Shapes and origins'. *Astrophysical Journal* 347, 29.

*Sanders, D.B., Scoville, N.Z. and Soifer, B.T. 1991. 'Molecular gas in luminous infrared galaxies'. *Astrophysical Journal* 370, 158.

*Sanders, D.B. and Mirabel, I.F. 1996. 'Luminous infrared galaxies'. *Annual Review of Astronomy and Astrophysics* 34, 749.

*Saunders, W. et al. 1990. 'The 60-micron and far-infrared luminosity functions of *IRAS* galaxies'. *Monthly Notices of the Royal Astronomical Society* 242, 318.

*Saunders, W. et al. 1991. 'The density field of the local universe'. *Nature* 349, 32.

*Saunders, W. et al. 2000.'The PSCz catalogue'. *Monthly Notices of the Royal Astronomical Society* 317, 55.

*Savage, B.D. and Mathis, J.S. 1979. 'Observed properties of interstellar dust'. *Annual Review of Astronomy and Astrophysics* 17, 73.

*Schlegel, D.J., Finkbeiner, D.P. and Davis, M. 1998. 'Maps of dust infrared emission for use in estimation of reddening and cosmic microwave background radiation foregrounds'. *Astrophysical Journal* 500, 525.

*Schödel, R. et al. 2002. 'A star in a 15.2-year orbit around the supermassive black hole at the centre of the Milky Way'. *Nature* 419, 694.

*Schödel, R. et al. 2003. 'Stellar dynamics in the central arcsecond of our galaxy'. *Astrophysical Journal* 596, 1015.

*Scott, S.E. et al. 2002. 'The SCUBA 8-mJy survey – I. Submillimetre maps, sources and number counts'. *Monthly Notices of the Royal Astronomical Society* 331, 817. 250

*Scoville, N.Z. and Solomon, P.M. 1974. 'Radiative transfer, excitation, and cooling of molecular emission lines (CO and CS)'. *Astrophysical Journal Letters* 187, L67.

Scoville, N.Z., Solomon, P.M. and Penzias, A.A. 1975. 'The molecular cloud Sagittarius B2'. *Astrophysical Journal* 201, 352.

*Scoville, N.Z. and Kwan, J. 1976. 'Infrared sources in molecular clouds'. *Astrophysical Journal* 206, 718.

*Scoville, N. and Young, J.S. 1983. 'The molecular gas distribution in M51'. *Astrophysical Journal* 265, 148.

*Scoville, N.Z. et al. 1987. 'Molecular clouds and cloud cores in the inner Galaxy'. *Astrophysical Journal Supplement Series* 63, 821.

*Scoville, N.Z., Sargent, A.I., Sanders, D.B. and Soifer, B.T. 1991. 'Dust and gas in the core of ARP 220 (IC 4553)'. *Astrophysical Journal Letters* 366, L5.

*Scoville, N.Z. et al. 2000. 'NICMOS imaging of infrared-luminous galaxies'. *Astronomical Journal* 119, 991.

Scoville, N. et al. 2007. 'The Cosmic Evolution Survey (COSMOS): overview'. *Astrophysical Journal Supplement Series* 172, 1.

Seebeck, T.J. 1826. 'Uber die magnetishe Polarisation der Metalle und Erze durch Temperatur-Differenz'. *Annalen der Physik* 82, 133.

*Sellgren, K. 1984. 'The near-infrared continuum emission of visual reflection nebulae'. *Astrophysical Journal* 277, 623.

Seyfert, K. 1943. 'Nuclear emission in spiral galaxies'. *Astrophysical Journal* 97, 28.

Shklovsky, I.S. 1952. 'The possibility of observing monochromatic radio emission from interstellar molecules'. *Astronomicheskii Zhurnal* 29, 144.

*Shu, F.H., Adams, F.C. and Lizano S. 1987. 'Star formation in molecular clouds – Observation and theory'. *Annual Review of Astronomy and Astrophysics* 25, 23.

*Shull, J.M. and Beckwith, S. 1982. 'Interstellar molecular hydrogen'. *Annual Review of Astronomy and Astrophysics* 20, 163.

*Silva, L., Granato, G.L., Bressan, A. and Danese, L. 1998. 'Modeling the effects of dust on Galactic spectral energy distributions from the ultraviolet to the millimeter band'. *Astrophysical Journal* 509, 103.

*Simon, M. et al. 1983. 'Infrared line and radio continuum emission of circumstellar ionized regions'. *Astrophysical Journal* 266, 623.

*Simon, M. et al. 1995. 'A lunar occultation and direct imaging survey of multiplicity in the Ophiuchus and Taurus star-forming regions'. *Astrophysical Journal* 443, 625.

*Skillman, E.D. and Kennicutt, R.C., Jr. 1993. 'Spatially resolved optical and near-infrared spectroscopy of I ZW 18'. *Astrophysical Journal* 411, 655.

*Skrutskie, M.F. et al. 2006. 'The Two Micron All Sky Survey (2MASS)'. *Astronomical Journal* 131, 1163.

*Smail, I., Ivison, R.J. and Blain, A.W. 1997. 'A deep sub-millimeter survey of lensing clusters: a new window on galaxy formation and evolution'. *Astrophysical Journal Letters* 490, L5.

*Smail, I., Ivison, R.J., Blain, A.W. and Kneib, J.-P. 2002. 'The nature of faint submillimetre-selected galaxies'. *Monthly Notices of the Royal Astronomical Society* 331, 495.

*Smith, B.A. and Terrile, R.J. 1984. 'A circumstellar disk around Beta Pictoris'. *Science* 226, 1421.

Smith, G.M. and Squibb, G.F. 1984. 'Development of the first infrared satellite observatory'. NASA Document 19840035078.

Smoot, G.F., Gorenstein, M.V. and Muller, R.A. 1977. 'Detection of anisotropy in the cosmic blackbody radiation'. *Physical Review Letters* 39, 898.

Smoot, G.F. et al. 1992. 'Structure on the COBE differential microwave radiometer first-year maps'. *Astrophysical Journal Letters* 396, L1.

Smoot, G. and Davidson, K. 1993. *Wrinkles in Time*. New York: William Morrow.

*Snell, R.L., Scoville, N.Z., Sanders, D.B. and Erickson, N.R. 1984. 'High-velocity molecular jets'. *Astrophysical Journal* 284, 176.

Snyder, L.E., Buhl, D., Zuckerman, B. and Palmer, P. 1969. 'Microwave detection of interstellar formaldehyde'. *Physical Review Letters* 22, 679.

Snyder, L.E. and Buhl, D. 1971. 'Observations of radio emission from interstellar hydrogen cyanide'. *Astrophysical Journal Letters* 163, L47.

Soifer, B.T., Houck, J.R. and Harwit, M. 1971. 'Rocket-infrared observations of the interplanetary medium'. *Astrophysical Journal Letters* 168, 73.

*Soifer, B.T. et al. 1984a. 'Infrared galaxies in the *IRAS* minisurvey'. *Astrophysical Journal Letters* 278, L71.

Soifer, B.T. et al. 1984b. 'The remarkable infrared galaxy Arp 220 = IC 4553'. *Astrophysical Journal Letters* 283, L1.

Soifer, B.T. et al. 1986. 'The luminosity function and space density of the most luminous galaxies in the IRAS survey'. *Astrophysical Journal Letters* 303, L41.

*Soifer, B.T. et al. 1987. 'The *IRAS* bright galaxy sample. II – The sample and luminosity function'. *Astrophysical Journal* 320, 238.

*Soifer, B.T., Neugebauer, G. and Houck, J.R. 1987. 'The *IRAS* view of the extragalactic sky'. *Annual Review of Astronomy and Astrophysics* 25, 187.

*Soifer, B.T., Boehmer, L., Neugebauer, G. and Sanders, D.B. 1989. 'The *IRAS* bright galaxy sample. IV – Complete *IRAS* observations'. *Astronomical Journal* 98, 766.

Soifer, B.T., Helou, G. and Werner, M. 2008. 'The Spitzer view of the extragalactic universe'. *Annual Review of Astronomy and Astrophysics* 46, 201.

*Solomon, P.M., Rivolo, A.R., Barrett, J. and Yahil, A. 1987. 'Mass, luminosity, and line width relations of Galactic molecular clouds'. *Astrophysical Journal* 319, 730.

*Solomon, P.M. and Sage, L.J. 1988. 'Star-formation rates, molecular clouds, and the origin of the far-infrared luminosity of isolated and interacting galaxies'. *Astrophysical Journal* 334, 613.

*Solomon, P.M., Downes, D. and Radford, S.J.E. 1992a. 'Dense molecular gas and starbursts in ultraluminous galaxies'. *Astrophysical Journal Letters* 387, L55.

Solomon, P.M., Downes, D. and Radford, S.J.E. 1992b. 'Warm molecular gas in the primeval galaxy IRAS F10214+4724'. *Astrophysical Journal Letters* 398, L29.

*Solomon, P.M., Downes, D., Radford, S.J.E. and Barrett, J.W. 1997. 'The molecular interstellar medium in ultraluminous infrared galaxies'. *Astrophysical Journal* 478, 144.

Solomon, P.M. and Vanden Bout, P.A. 2005. 'Molecular gas at high redshift'. *Annual Review of Astronomy and Astrophysics* 43, 677.

*Sopka, R.J. et al. 1985. 'Submillimeter observations of evolved stars'. *Astrophysical Journal* 294, 242.

Spergel, D.N. et al. 2003. 'First-year Wilkinson Microwave Anisotropy Probe (WMAP) observations: determination of cosmological parameters'. *Astrophysical Journal Supplement Series* 148, 175.

Spergel, D.N. et al. 2007. 'Three-year Wilkinson Microwave Anisotropy Probe (WMAP) observations: implications for cosmology'. *Astrophysical Journal Supplement Series* 170, 377.

*Spitzer, L. 1978. *Physical Processes in the Interstellar Medium*. New York: Wiley.

Spoon, H.W.W. et al. 2006. 'The detection of crystalline silicates in ultraluminous infrared galaxies'. *Astrophysical Journal* 638, 759.

*Stahler, S.W., Shu, F.H. and Taam, R.E. 1980. 'The evolution of protostars. I – Global formulation and results'. *Astrophysical Journal* 241, 148.

Stebbins, J., Huffer, C.M. and Whitford, A.E., 1939. 'The mean coefficient of selective absorption in the Galaxy'. *Astrophysical Journal* 92, 193.

*Stecker, F.W., de Jager, O.C. and Salamon, M.H. 1992. 'TeV gamma rays from 3C 279 – A possible probe of origin and intergalactic infrared radiation fields'. *Astrophysical Journal Letters* 390, L49.

Stein, W.A. 1966a. 'Infrared emission from interstellar grains'. *Astrophysical Journal* 144, 318.

Stein, W.A. 1966b. 'Infrared emission by circumstellar dust'. *Astrophysical Journal* 145, 101.

Stein, W.A. 1967. 'Infrared continuum for HII regions'. *Astrophysical Journal* 148, 295.

*Stern, D. et al. 2005. 'Mid-infrared selection of active galaxies'. *Astrophysical Journal* 631, 163.

*Sternberg, A. and Dalgarno, A. 1989. 'The infrared response of molecular hydrogen to ultraviolet radiation – High-density regions'. *Astrophysical Journal* 338, 197.

Stone, E.F. 1870. 'Approximate determinations of the heating powers of Arcturus and Alpha Lyrae'. *Proceedings of the Royal Society of London* 18, 159.

Storey, J.W.V. 2000. 'Infrared astronomy: In the heat of the night. The 1999 Ellery Lecture'. *Publications of the Astronomical Society of Australia* 17, 270.

*Strauss, M.A. et al. 1992. 'A redshift survey of *IRAS* galaxies. VII – The infrared and redshift data for the 1.936 Jansky sample'. *Astrophysical Journal Supplement Series* 83, 29.

Strecker, D.W., Ney, E.P. and Murdock, T.L. 1973. 'Cygnids and Taurids – Two classes of infrared objects'. *Astrophysical Journal Letters* 183, L13.

*Strom, K.M. et al. 1989. 'Circumstellar material associated with solar-type pre-main-sequence stars – A possible constraint on the timescale for planet building'. *Astronomical Journal* 97, 1451.

Strom, R.G. et al. 2005. 'The origin of planetary impactors in the inner Solar System'. *Science* 309, 1847.

*Strom, S.E., Grasdalen, G.L. and Strom, K.M. 1974. 'Infrared and optical observations of Herbig–Haro objects'. *Astrophysical Journal* 191, 111.

*Stutzki, J. et al. 1988. 'Submillimeter and far-infrared line observations of M17 SW – A clumpy molecular cloud penetrated by ultraviolet radiation'. *Astrophysical Journal* 332, 379.

Swain, M.R., Vasisht, G. and Tinetti, G. 2008. 'The presence of methane in the atmosphere of an extrasolar planet'. *Nature* 452, 329.

Sykes, M.V., Lebofsky, L.A., Hunten, D.M. and Low, F. 1986. 'The discovery of dust trails in the orbits of periodic comets'. *Science* 232, 1115.

Sykes, M.V. and Walker, R.G. 1992. 'Cometary dust trails. I – Survey'. *Icarus* 95, 180.

Tacconi, L.J. et al. 2006. 'High-resolution millimetre imaging of submillimeter galaxies'. *Astrophysical Journal* 640, 228.

Tatematus, K. et al. 1993. 'Nobeyama CS(1–0) Survey'. *Astrophysical Journal* 404, 643.

Tauber, J.A. et al. 2010. 'Planck pre-launch status: The Planck mission'. *Astronomy and Astrophysics* 520, A1.

Tedesco, E.F., Noah, P.V., Noah, M. and Price, S.D. 2002. 'The Supplemental IRAS Minor Planet Survey'. *Astronomical Journal* 123, 1056.

Telesco, C.M., Harper, D.A. and Loewenstein, R.F. 1976. 'Far-infrared photometry of NGC 1068'. *Astrophysical Journal Letters* 203, L53.

*Telesco, C.M. and Harper, D.A. 1980. 'Galaxies and far-infrared emission'. *Astrophysical Journal* 235, 392.

Thompson, R.I. et al. 1998. 'Initial on-orbit performance of NICMOS'. *Astrophysical Journal Letters* 492, L95.

*Tielens, A.G.G.M. and Hagen, W. 1982. 'Model calculations of the molecular composition of interstellar grain mantles'. *Astronomy and Astrophysics* 114, 245.

*Tielens, A.G.G.M. and Hollenbach, D. 1985. 'Photodissociation regions. I – Basic model. II – A model for the Orion photodissociation region'. *Astrophysical Journal* 291, 722.

*Tielens, A.G.G.M., Tokunaga, A.T., Geballe, T.R. and Baas, F. 1991. 'Interstellar solid CO – Polar and nonpolar interstellar ices'. *Astrophysical Journal* 381, 181.

Tinetti, G. et al. 2007. 'Water vapour in the atmosphere of a transiting extrasolar planet'. *Nature* 448, 169.

*Tonry, J.L. et al. 2003. 'Cosmological results from high-*z* supernovae'. *Astrophysical Journal* 594, 1.

*Townes, C.H. and Schawlow, A.L. 1955. *Microwave Spectroscopy.* New York: McGraw-Hill.

*Trumpler, R.J. 1930. 'Preliminary results on the distances, dimensions and space distribution of open star clusters'. *Lick Observatory Bulletin* 14, 154.

Tsiganis, K., Gomes, R., Morbidelli, A. and Levison, H.F. 2005. 'Origin of the orbital architecture of the giant planets in the Solar System'. *Nature* 435, 459.

Ulrich, B.T. et al. 1966. 'Further observations of extremely cool stars'. *Astrophysical Journal* 146, 288.

*Ungerechts, H. and Thaddeus, P. 1987. 'A CO survey of the dark nebulae in Perseus, Taurus, and Auriga'. *Astrophysical Journal Supplement Series* 63, 645.

*van der Veen, W.E.C.J. and Habing, H.J. 1988. 'The *IRAS* two-colour diagram as a tool for studying late stages of stellar evolution'. *Astronomy and Astrophysics* 194, 125.

*van Dishoeck, E.F. and Black, J.H. 1986. 'Comprehensive models of diffuse interstellar clouds – Physical conditions and molecular abundances'. *Astrophysical Journal Supplement Series* 62, 109.

*van Dishoeck, E.F. and Blake, G.A. 1998. 'Chemical evolution of star-forming regions'. *Annual Review of Astronomy and Astrophysics* 36, 317.

van Dishoeck, E.F. 2004. 'ISO spectroscopy of gas and dust: from molecular clouds to protoplanetary disks'. *Annual Review of Astronomy and Astrophysics* 42, 119.

*Veeder, G.J. 1974. 'Luminosities and temperatures of M dwarf stars from infrared photometry'. *Astronomical Journal* 79, 1056.

*Veilleux, S. et al. 1995. 'Optical spectroscopy of luminous infrared galaxies. II. Analysis of the nuclear and long-slit data'. *Astrophysical Journal Supplement Series* 98, 171.

Wagoner, R.V., Fowler, W.A. and Hoyle, F. 1967. 'On the synthesis of elements at very high temperatures'. *Astrophysical Journal* 148, 3.

*Wainscoat, R.J. et al. 1992. 'A model of the 8–25 micron point source infrared sky'. *Astrophysical Journal Supplement Series* 83, 111.

*Wainscoat, R.J. and Cowie, L.L. 1992. 'A filter for deep near-infrared imaging'. *Astronomical Journal* 103, 332.

Walker, R.G. and Price, S.D. 1975. *AFCRL Infrared Sky Survey.* Cambridge, MA: Air Force Cambridge Research Laboratory (AFCRL-TR-75-0373).

Walker, R.G. et al. 1984. 'Observations of Comet IRAS-Iraki-Alcock 1983d'. *Astrophysical Journal Letters* 278, L11.

Walker, R.G. and Rowan-Robinson, M. 1984. 'The peculiar infrared tail of Comet Bowell'. *Bulletin of the American Astronomical Society* 16, 443.

*Ward-Thompson, D., Scott, P.F., Hills, R.E. and André, P. 1994. 'A submillimetre continuum survey of pre-protostellar cores'. *Monthly Notices of the Royal Astronomical Society* 268, 276.

Ward-Thompson, D., Motte, F. and Andre, P. 1999. 'Mapping of pre-stellar cores'. *Monthly Notices of the Royal Astronomical Society* 305, 143.

*Warren, S.G. 1984. 'Optical constants of ice from the ultraviolet to the microwave'. *Applied Optics* 23, 1206.

Warren, S.J. et al. 2007a. 'The UKIRT Deep Sky Survey first data release'. *Monthly Notices of the Royal Astronomical Society* 375, 213.

Warren, S.J. et al. 2007b. 'A very cool brown dwarf in UKIDSS DR1'. *Monthly Notices of the Royal Astronomical Society* 381, 1400.

*Waters, L.B.F.M. et al. 1996. 'Mineralogy of oxygen-rich dust shells'. *Astronomy and Astrophysics* 315, L361.

*Webb, R.A. et al. 1999. 'Discovery of seven T Tauri stars and a brown dwarf candidate in the nearby TW Hydrae Association'. *Astrophysical Journal Letters* 512, L63.

Weinreb, S., Barrett, A.H., Meeks, M.L., Henry, J.C. 1963. 'Radio observations of OH in the interstellar medium'. *Nature* 200, 829.

*Weingartner, J.C. and Draine, B.T. 2001. 'Dust grain-size distributions and extinction in the Milky Way, Large Magellanic Cloud, and Small Magellanic Cloud'. *Astrophysical Journal* 548, 296.

Weliachew, L. 1971. 'Detection of interstellar OH in two external galaxies'. *Astrophysical Journal Letters* 147, L47.

Werner, M.W., Elias, J.H., Gezari, D.Y., Hauser, M.G. and Westbrook, W.E. 1975. 'Observations of 1-millimeter continuum radiation from the DR 21 region'. *Astrophysical Journal Letters* 199, L185.

*Werner, M.W. et al. 2004. 'The Spitzer Space Telescope mission'. *Astrophysical Journal Supplement Series* 154, 1.

Werner, M. 2006. 'A short and personal history of the Spitzer Space Telescope'. *Astronomical Society of the Pacific Conference Series* 357, 7

Werner, M. et al. 2006. 'First fruits of the Spitzer Space Telescope: Galactic and Solar System studies'. *Annual Review of Astronomy and Astrophysics* 44, 269.

Wesselink, A.H. 1948. 'Heat conductivity and nature of the lunar surface material'. *Bulletin of the Astronomical Institutes of the Netherlands* 10, 351.

*Whitford, A.E. 1958. 'The law of interstellar reddening'. *Astronomical Journal* 63, 201.

*Whittet, D.C.B. et al. 1988. 'Infrared spectroscopy of dust in the Taurus dark clouds – Ice and silicates'. *Monthly Notices of the Royal Astronomical Society* 233, 321.

*Wilking, B.A. and Lada, C.J. 1983. 'The discovery of new embedded sources in the centrally condensed core of the Rho Ophiuchi dark cloud – The formation of a bound cluster'. *Astrophysical Journal* 274, 698.

*Wilking, B.A., Lada, C.J. and Young, E.T. 1989. 'IRAS observations of the Rho Ophiuchi infrared cluster – Spectral energy distributions and luminosity function'. *Astrophysical Journal* 340, 823.

*Williams, J.P., de Geus, E.J. and Blitz, L. 1994. 'Determining structure in molecular clouds'. *Astrophysical Journal* 428, 693.

*Williams, P.M. et al. 1990. 'Multi-frequency variations of the Wolf–Rayet system HD 193793. I – Infrared, X-ray and radio observations'. *Monthly Notices of the Royal Astronomical Society* 243, 662.

*Willner, S.P. et al. 1982. 'Infrared spectra of protostars – Composition of the dust shells'. *Astrophysical Journal* 253, 174.

Wilson, R.W., Jefferts, K.B. and Penzias, A.A. 1970. 'Carbon monoxide in the Orion Nebula'. *Astrophysical Journal Letters* 161, L43.

Wilson, R.W., Solomon, P.M., Penzias, A.A. and Jefferts, K.B. 1971. 'Millimeter observations of CO, CN, and CS emission from IRC+10216'. *Astrophysical Journal Letters* 169, L35.

*Wilson, W.J. et al. 1972. 'Infrared stars with strong 1665/1667-MHz OH microwave emission'. *Astrophysical Journal* 177, 523.

Wooden, D.H. et al. 1999. 'Silicate mineralogy of the dust in the inner coma of comet C/1995 01 (Hale-Bopp) pre- and postperihelion'. *Astrophysical Journal* 517, 1034.

Woody, D.P., Mather, J.C., Nishioka, N.S. and Richards, P.L. 1975. 'Measurement of the spectrum of the submillimeter cosmic background'. *Physical Review Letters* 34, 1036.

Woody, D.P. and Richards, P.L. 1979. 'Spectrum of the cosmic background radiation'. *Physical Review Letters* 42, 925.

*Woolf, N. and Ney, E. 1969. 'Circumstellar emission from cool stars'. *Astrophysical Journal Letters* 155, L181.

*Wright, A.E. and Barlow, M.J. 1975. 'The radio and infrared spectrum of early-type stars undergoing mass-loss'. *Monthly Notices of the Royal Astronomical Society* 170, 41.

*Wright, E.L. et al. 1991. 'Preliminary spectral observations of the Galaxy with a 7 deg beam by the Cosmic Background Explorer (COBE)'. *Astrophysical Journal* 381, 200.

Wright, G.S., Joseph, R.D. and Meikle, W.P.S. 1984. 'The ultraluminous interacting galaxy NGC 6240'. *Nature* 309, 430.

Wyatt, M.C. 2008. 'Evolution of debris disks'. *Annual Review of Astronomy and Astrophysics* 46, 339.

*Wynn-Williams, C.G., Becklin, E.E. and Neugebauer, G. 1972. 'Infra-red sources in the H II region W3'. *Monthly Notices of the Royal Astronomical Society* 160, 1.

*Wynn-Williams, C.G., Becklin, E.E. and Neugebauer, G. 1974. 'Infrared studies of H II regions and OH sources'. *Astrophysical Journal* 187, 473.

Wynn-Williams, C.G. and Becklin, E.E. (eds.). 1987. *Astronomy with Infrared Arrays*. Honolulu: University of Hawaii.

Yahil, A., Walker, D. and Rowan-Robinson, M. 1986. 'The dipole anisotropies of the IRAS galaxies and the microwave background radiation'. *Astrophysical Journal Letters* 301, L1.

Yorke, H. 1977. 'Calculated infrared spectra of cocoon stars'. '*Astronomy and Astrophysics* 58, 423.

*Young, J.S. and Scoville, N. 1982. 'Extragalactic CO – Gas distributions which follow the light in IC 342 and NGC 6946'. *Astrophysical Journal* 258, 467.

*Young, J.S., Xie, S., Kenney, J.D.P. and Rice, W.L. 1989. 'Global properties of infrared bright galaxies'. *Astrophysical Journal Supplement Series* 70, 699.

*Young, J.S. and Scoville, N.Z. 1991. 'Molecular gas in galaxies'. *Annual Review of Astronomy and Astrophysics* 29, 581.

*Young, J.S. et al. 1995. 'The FCRAO extragalactic CO survey. I. The data'. *Astrophysical Journal Supplement Series* 98, 219.

*Yun, M.S., Reddy, N.A. and Condon, J.J. 2001. 'Radio properties of infrared-selected galaxies in the IRAS 2 Jy sample'. *Astrophysical Journal* 554, 803.

*Zuckerman, B. 1973. 'A model of the Orion Nebula'. *Astrophysical Journal* 183, 863.

*Zuckerman, B. and Palmer, P. 1974. 'Radio radiation from interstellar molecules'. *Annual Review of Astronomy and Astrophysics* 12, 279.

*Zuckerman, B. and Aller, L.H. 1986. 'Origin of planetary nebulae – Morphology, carbon-to-oxygen abundance ratios, and central star multiplicity'. *Astrophysical Journal* 301, 772.

Name Index

Subject Index